This book is intended to help those students who have undertaken medical biochemistry, whether you are a medical student or other health professional. It is intended to provide a concise understanding of the main points of medical biochemistry to assist you in understanding the broader concepts that you will apply through your education.

This book will be of value to those students interested in a concise overview of Medical Biochemistry.
The manner that the information is presented for the student will provide a simplified description of complex concepts.

Perry Hudson M.D.

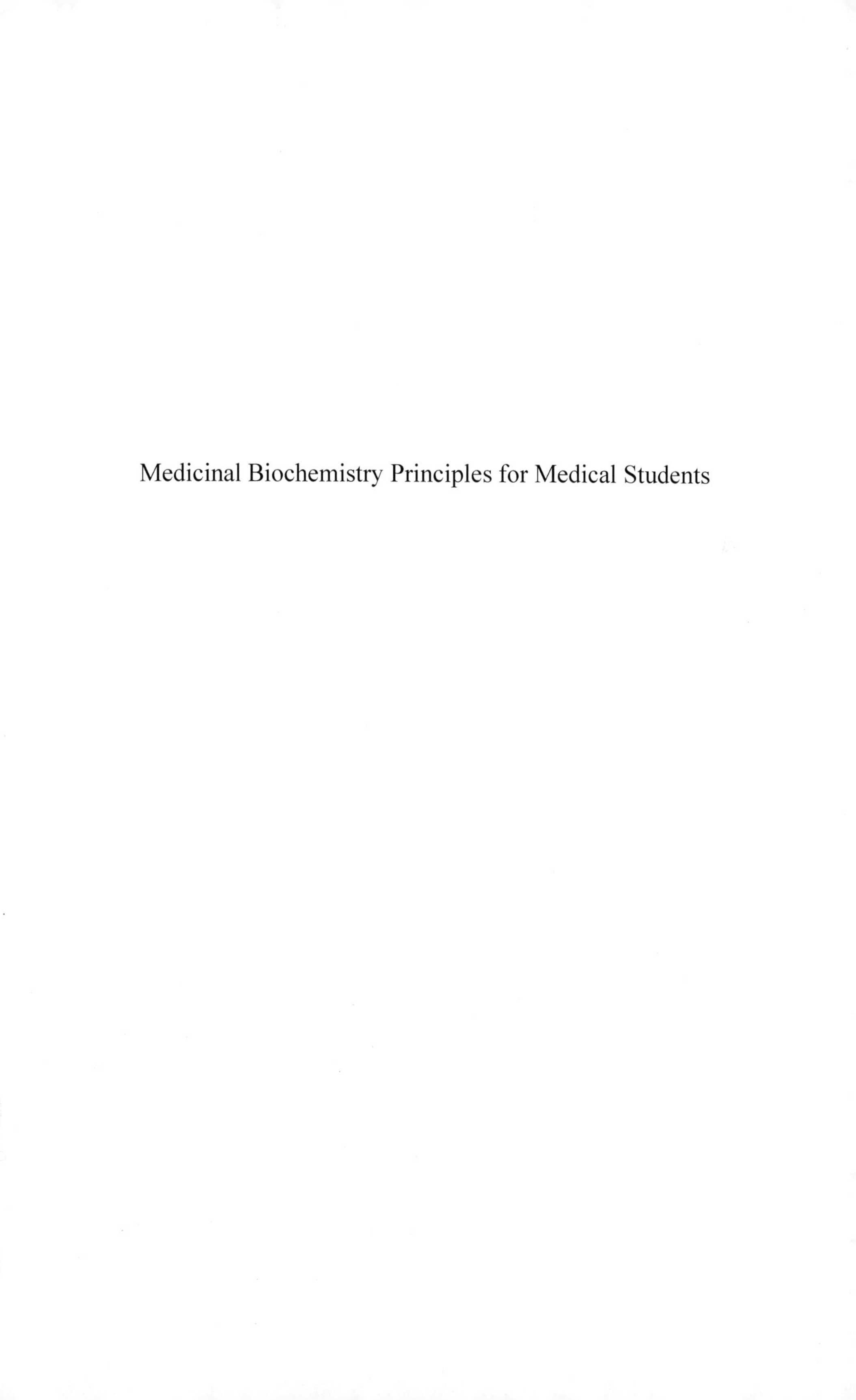

Medicinal Biochemistry Principles for Medical Students

Other books written by the same author include:

- Gross Anatomy of the Head & Neck
- Neuroscience for the Health Professional
- Concussive wave blast associated brain tissue damage

Medical Biochemistry Principles for Medical Students

David W. Karam M.D., Ph.D

Order this book online at www.trafford.com
or email orders@trafford.com

Most Trafford titles are also available at major online book retailers.

© Copyright 2011 David W. Karam M.D., Ph.D.
All rights reserved. No part of this publication may be reproduced, stored in a retrieval system, or transmitted, in any form or by any means, electronic, mechanical, photocopying, recording, or otherwise, without the written prior permission of the author.

Printed in the United States of America.

ISBN: 978-1-4269-5873-1 (sc)
ISBN: 978-1-4269-5874-8 (e)

Trafford rev. 02/28/2011

www.trafford.com

North America & International
toll-free: 1 888 232 4444 (USA & Canada)
phone: 250 383 6864 • fax: 812 355 4082

"For all those who enlightened me,
scolded me and encouraged me."

"If it looks like a horse, walks like a horse,
whinny's like a horse, don't think it's a Zebra"

Dr. Paul Lennard
Professor Emory University

Biochemistry is one of those subjects that students either love or hate, yet is such an integral part of medicine that students spend many laborious hours trying to memorize the basic principles. In the many years of training medical students, interns, residents and fellows, I refer them back to the basic principles of biochemistry as the building blocks to all else that makes up the human body. This text will make clear to the reader those principles that are important to understand how we as humans function in a concise and clear format.

Perry Hudson M.D.

I hope that this work will provide a concise yet thorough description of common principles of medical biochemistry.

CONTENTS

Chemistry Review

I. Matter – Occupies space & has mass

   a. Mass is intrinsic property of matter

   b. Weight is the gravitational attraction on an object

   c. Divided into : distinct substances & mixtures.

      i. Distinct substances are elements or compounds. The element is a single kind of atom, while the compound is two or more kinds of atoms joined in a definitive manner

      ii. Mixture - two or more distinct substances that are intimately joined with no defining properties. Homogenous mixture – the composition of the compounds are uniform throughout.

II. Force- related to mass through Newton's equation $F=ma$. Force is equal to mass times acceleration.

   a. Weight and mass are related through the equation $W=mg$. Weight is the mass times the acceleration of gravity

III. Properties of matter – States or phases of matter – solid/ liquid /gas

   a. Solid – defined size and shape

   b. Liquid – defined volume

   c. Gas – neither of the above

d. Extensive properties of matter – mass & volume – depend on the size of the sample

e. Intensive properties of matter – Melting point Tm, boiling point, density are independent on the size of the sample.

f. Physical properties of matter – observed - physical state, color, Tm.

g. Chemical changes – involve breaking or forming or altering chemical bonds

h. Physical changes – do not result in the production of a new substance.

i. Conservation of matter – Matter is neither created or destroyed, it only changes from one form to another

IV. Energy – the ability to do work or produce heat. Energy is available in many forms, light, sound, heat , mechanical, electrical & chemical. Energy can be transferred from one form to another.

a. Energy classifications

i. Potential Energy – energy of position, energy in waiting so to speak.

ii. Kinetic Energy – energy of motion and is proven by the formula $KE=1/2\ ms^2$ (half of the objects mass times the speed squared)

b. Conservation of Energy – is the theory energy cannot be created or destroyed, only changed from one form to another.

c. Entropy is the measure of randomness that a system represents, therefore the greater the randomness, the greater the entropy. The entropy of the universe increases for any spontaneous process:

$\Delta S_{universe} = (\Delta S_{system} + \Delta S_{surroundings}) \geq 0.$

I. The lowest energy state of the atom is the Ground state. Any increase in the energy status of the atom is its excited state. The formula for the definition of the differences of the two are:

a. Represented by the $\Delta E = E_{final} - E_{initial.}$

b. As the electron goes from the ground state to the excited state it will absorb energy

c. As the electron goes from the excited state to the ground state it will give off energy

II. Periodic Table arrangement

Fig 1 The Periodic Table

1 1A	2 2A	3	4	5	6	7	8	9	10	11	12	13 3A	14 4A	15 5A	16 6A	17 7A	18 8A
Alkali metals	Alkaline earth metals	Transition metals														Halogens	Noble gases
1 H 1.008																	2 He 4.003
3 Li 6.941	4 Be 9.012											5 B 10.81	6 C 12.01	7 N 14.01	8 O 16.00	9 F 19.00	10 Ne 20.18
11 Na 22.99	12 Mg 24.31											13 Al 26.98	14 Si 28.09	15 P 30.97	16 S 32.07	17 Cl 35.45	18 Ar 39.95
19 K 39.10	20 Ca 40.08	21 Sc 44.96	22 Ti 47.88	23 V 50.94	24 Cr 52.00	25 Mn 54.94	26 Fe 55.85	27 Co 58.93	28 Ni 58.69	29 Cu 63.55	30 Zn 65.38	31 Ga 69.72	32 Ge 72.59	33 As 74.92	34 Se 78.96	35 Br 79.90	36 Kr 83.80
37 Rb 85.47	38 Sr 87.62	39 Y 88.91	40 Zr 91.22	41 Nb 92.91	42 Mo 95.94	43 Tc (98)	44 Ru 101.1	45 Rh 102.9	46 Pd 106.4	47 Ag 107.9	48 Cd 112.4	49 In 114.8	50 Sn 118.7	51 Sb 121.8	52 Te 127.6	53 I 126.9	54 Xe 131.3
55 Cs 132.9	56 Ba 137.3	57 La* 138.9	72 Hf 178.5	73 Ta 180.9	74 W 183.9	75 Re 186.2	76 Os 190.2	77 Ir 192.2	78 Pt 195.1	79 Au 197.0	80 Hg 200.6	81 Tl 204.4	82 Pb 207.2	83 Bi 209.0	84 Po (209)	85 At (210)	86 Rn (222)
87 Fr (223)	88 Ra 226	89 Ac† (227)	104 Unq	105 Unp	106 Unh	107 Uns	108 Uno	109 Une									

metals / nonmetals

*Lanthanides	58 Ce 140.1	59 Pr 140.9	60 Nd 144.2	61 Pm (145)	62 Sm 150.4	63 Eu 152.0	64 Gd 157.3	65 Tb 158.9	66 Dy 162.5	67 Ho 164.9	68 Er 167.3	69 Tm 168.9	70 Yb 173.0	71 Lu 175.0
†Actinides	90 Th 232.0	91 Pa (231)	92 U 238.0	93 Np (237)	94 Pu (244)	95 Am (243)	96 Cm (247)	97 Bk (247)	98 Cf (251)	99 Es (252)	100 Fm (257)	101 Md (258)	102 No (259)	103 Lr (260)

a. Vertical columns are groups, with each group representing elements having like chemical properties

b. Horizontal rows are called periods. Those elements in the 2 rows below the main body of the periodic table are the inner transition elements.
   i. Top row (elements 58-71) are the lanthanides or the rare earths
   ii. Bottom row (elements 90-103) are the actinides
c. Group IA elements are the alkali metals
d. Group IIA elements are the alkaline earth metals
e. Group VIIA are the halogens
f. Group O elements are the noble gases.
g. Groups IB thru VIIB & Group VIII are the transition elements.

III. The most active metals are known to be in the lower left corner.

IV. The most active nonmetals are found in the upper right
   a. Metallic properties are increased electrical conductivity, luster, high Tm, malleability.
   b. Properties also include reactivity with acids

V. Non metals are poor electrical conductors and form very brittle solids

VI. Metalloid properties will fall between the above

VII. Atomic Radius will decrease across the period from L→R, and will increase down a group.

VIII. Electronegativity measures the strength with which the atoms of the element will attract valence electrons for chemical bonding.

IX. Ionization Energy is the energy required to remove the electron from an isolated atom in its grounded state. This value will decrease as you go down a group, but will increase as you go across L-→R

X. Acid forming properties of elements will increase as you move L -→R .

XI. The Bohr Theory states that an electron can exist only in certain stable energy levels, and altering these levels the amount of energy change must be equal to the exact difference between the final and the initial state.

a. Eq: $\Delta E = Ea - Eb$

b. $Eb - Ea = z^2e^2/2a_0 \left[\frac{1}{na2} - \frac{1}{nb2}\right]$ This will measure the energy difference between the states a & b, where n= to the quantum energy level, E = to the energy, e = to the charge on the electron, a_0 = to the Bohr radius and z = atomic number.

XII. Components of the Atomic Structure

a. Mass number – the number of protons & neutrons. This number corresponds to the isotopic atomic weight.

b. Atomic number - the number of protons within the nucleus.

c. Valence electrons – the electrons found in the outermost shell. These are important for the oxidation state of the atom, when electrons are lost or partially lost (via sharing) there is a positive (+) oxidative number assigned. If the valence electrons are gained or partially gained, the oxidative number is assigned a negative (-) number. The Lewis Dot Structure is the diagrammatic representation of a molecules valence electrons for each atom. It is important that the formation of a stable compound be obtained from the configuration of the atoms towards that of a noble gas. In this representation, the most important thing is the sum total of the electrons and not those individually. This is due to the sharing of the bound electrons by the atoms within the compound.

d. The ground state is considered to the lowest energy state that the atom is able to attain. Therefore, any energy state that is above this state is considered to be a state of excitement. When the atom changes from the ground state to the excited state the atom must absorb energy. If this is considered factual, then the reverse must hold true, that if the atom is going from a state of excitability to the ground state the atom must loose energy.

XIII. The Pauli Exclusion principle means that no two electrons within the same atom can have the same four quantum numbers

a. Hunds Rule expresses the relationship of the electrons in each orbital. It states that for a set of equal energy orbitals, each orbital is occupied by one electron before any orbital has two.

b. Heisenberg Uncertainty Principle states that there is limited ability to accurately predict the exact position of a particle as it is in its path around in the valence shell.

c. The Aufbau Principle states that protons are added in a specific order and that electrons are added in exact conjunction.

XIV. Bonds between the elements or compounds are of a variety of classifications: ionic, covalent, non-polar covalent and polar covalent.

A. Ionic bonds are produced when one or more electrons of the valence shell are transferred from one atom to another atom's valence shell. This process will yields one positive cation and one negative anion. The principle that makes this occur is the columb attraction between the negative and the positive charges on the atoms. The octet rule comes into play with this loss or net gain of electron to the outer valence shell. The octet rule states that atoms will have the predisposition to gain or loose electrons from the outer valence shell until there are 8 electrons occupying that shell with those atoms that possess the 2s and the three 2p orbitals. In the description of the orbitals, it is imperative to understand the Lewis Structure. In writing the structure, the concern is the total number of electrons available. Take the electrons and create the bonds for each pair of bound atoms and then arrange the electrons that are remaining in such a way to comply with the duet rule for hydrogen and the octet rule for the 2nd row elements. (Boron is the exception to the octet rule as are the 3rd period elements as they are able to exceed the octet rule) The formal charge of an atom in a molecule is defined as the difference between the number of valence electrons on the free atom and the number of electrons assigned to the atom in the molecule. Lone pair electrons are assigned to the atom while shared electrons are divided equally between the sharing atoms. The total of the formal charges on an ion or molecule must equal the overall charge on the ion of the species.

B. Covalent bonds result form the sharing of a pair of valence electrons in the outer shell between two atoms. In non polar covalent bonds, these outer shell valence electrons are shared equally, where in the polar (dipole) covalent bonds, they are not equally shared. Homonuclear diatomic molecules are more predisposed to the formation of nonpolar covalent bonds.

C. Intermolecular Forces are described by the dipole distance that separates the positive and the negative charges. There is a mathematical definition of this attraction: it is the charge multiplied by the distance between the two charges = the dipole moment. The dipole moment will dictate the dipole force, the attraction of the oppositely charged substances. There are some very peculiar aspects to this type of bond especially when hydrogen atoms are one of the atoms in the bonding. If a hydrogen atom is bonding to a very electronegative atom, the hydrogen atom will take on a partial positive charge and increase the bonding of adjacent electron pairs. This increased attractive force will generate a hydrogen bond. The strength of the hydrogen bond is dependent upon the polarity of the atoms, the more polar, the greater the attraction and the stronger the bonding. If the attraction is weak it is called Van der Waals forces. The Van der Waals forces are apparent at low temperatures and increased pressures in most instances. These forces are good for the production of a soft crystalline structure. They are also easily deformed and vaporized due to the lower attraction of the bond type.

D. Double Vs. Triple Bonds are produced through the sharing of either two or three pairs of electrons. There is increased amounts of energy required in the liberation of double and triple bonds than seen in the liberation of a single bond. In addition to the increased energy associated with the double and triple bonds, there is an associated change in the interatomic distances within these bonds. The double and triple bonding will shorten the interatomic distances such that it is a contributing factor in the increased strength of these bonds.

E. VESPER - Valence Shell Electron Repulsion theory is the basis for the variable bond angles that can be formed. It is the theory that allows for the prediction of the bond just from the number of bonding and the non bonding electron pairs of the valence shell of the central atom.

Fig. 2

Number of Bonds	Number of unused e- Pairs	Hybrid Orbital	Bond Angle	Geometry	Example
2	0	sp	180	Linear	BeF2
3	0	sp^2	120	Trigonal Planar	BF3
4	0	sp^3	109.5	Tetrahedral	CH4
3	1	sp^3	90 - 109.5	Pyramidal	NH3
2	2	sp^3	90- 109.5	angular	H2O
6	0	sp^3d^2	90	Octahedral	SF6

F. Sigma and Pi Bonds are formed by the position of the electrons in the bond in relation to the nuclei. If the orbital is passing through the two nuclei it will be termed a sigma bond. If the orbital that is produced is found above and below the nuclei then it is termed a pi bond. Pi bonds are seen in the double and triple bonds seen in atomic compounds.

G. Ionic substances are brittle in nature. The ionic substances are also seen to form electrolytes if they are soluble in water. (Definition of electrolyte is that it is a good conductor of electricity), while they are poor conductors of electrical charge in their solid phase. In the crystal structures that ionic substances can form, the lattice energy is high owing to the very strong electrostatic attraction between the atoms. This is also a factor in the high melting (Tm) and boiling points associated with ionic substances.

H. In comparison to the ionic substances mentioned above, the molecular crystals have lower lattice energies, and both the solid and the liquid phases are poor conductors of electrical currents. The molecular crystals are able to exist as a gas at room temperature and atmospheric pressure. The melting and boiling points are relatively low as compared to the ionic substances.

Math of Chemistry

I. The mole is a value equal to the amount of substance containing Avogadro's number of particles (6.02×10^{23}) The formula is number of moles = $^{mass}/_{molecular\ weight}$

II. Atomic weight is the gram – atomic weight and is measured in grams. This value contains one mole of atoms of the element.

III. Molecular weight is also known as the formula weight. This is a derived value. It is the weight of the addition of the component atomic weights.

a. FW of $CaCo_3$ = 1(40) + 1(12) + 3(16) = 100 g/mol

b. Molecular weight : (density) x(volume per molecule) x(Avogadro's #) = mass of one mole of the molecule.

c. (Density) x(volume per mole) = mass of one molecule

IV. Density $= \frac{Mass}{Volume} \quad \frac{grams}{ml}$

V. Theoretical yield is the yield of the reaction to product if the reaction runs to completion rather than equilibrium.

VI. Percent yield is derived by the formula: $\frac{actual\ yield}{theoritical\ yield}$ x 100% This defines the efficiency of the reaction.

VII. Balancing a chemical equation results in the same number of atoms on each side of the reaction arrow. In the example of ($2 NaOH + H_2SO_4 \rightarrow Na_2 SO_4 + 2 H_2 O$), Na has 2 atoms, O has 6 atoms, H has 4 atoms, S has 1 atom in this accounting format. This is not able to be altered in the ability to balance chemical equations.

VIII. Coefficients in the chemical equations indicate the ratios of one mole of reactant that will react with moles of another reactant. In the reaction $C_2H_4 + 3O_2 \rightarrow 2 CO_2 + 2H_2O$ the coefficients indicate that 1 mole of C_2H_4 + 3 moles of O_2 $\rightarrow$ 2 moles CO_2 + 2 moles H_2O.

VIV. Empirical Formula VS. Molecular Formulas.

a. The molecular formula of any substance is indicative of the actual number of atoms in the molecule. This is a formula that is not easily derived. One must first calculate the empirical formula then the next step is to calculate the molecular formula by using the molecular weight of the substance. The basis of the molecular formula is as a whole number multiple of the empirical formula. In writing the final forms of a substance or compound, the subscript coefficients are always reduced the lowest common terms.

b. The empirical formula is the description of the relative number of atoms that are present in the substance. This represents the most basic descriptive form of the substance.

c. Compound names of binary compounds will show the two elements present with the defining ending of

"-ide" In this type of example, there are two states of the metal with regards to the oxidation states, the higher state being "-ic" and the lower state being "-ous". If the binary compound is a non metal, the Greek numbering system is used. If the compound is a tertiary compound then the configuration is more than likely going to be made up of an element and then a radical. In this situation, the positive element is named first then the negatively named one follows.

X. Length : 1 meter = 39.7 inches 1 meter = 100 cm 1 meter = 1000 mm 1 cm = 2.54 inches

A. Rectangular Volume: V= height x length x width

B. Spherical Volume : $V = \pi r^3$

Cylinder Volume: $V = \pi r^2 l$ 1 Liter = 1000 cc

C. Temperature: °C = 5/9(°F-32°) Kelvin °K = °C + 273.15

D. In multiplication of exponents the exponents are added, in Division they are subtracted.

XI. Solutions. There are three main types of solutions, the gaseous, the solid and the liquid. In the discussion of solutions, the most common type of solution is where there is a substance (solute) dissolved in a liquid.

A. solids as a solution are demonstrated by the alloys

B. The gaseous solution is classically demonstrated by the air we breath.

C. when a solution is produced there is a solvation reaction that is carried out. In this type of reaction, there will be an interaction of the solvent molecules with the solute molecules. This reaction will produce loosely bonded conglomeration molecules. Water is often used as the solvent and when used the reaction is also called "hydration reaction"

D. Solutions with the solute - solute, solute - solvent, and the solvent - solvent interactions are all the same are called ideal solutions.

E. The solubility of most solids in liquids will increase as the solution is heated, (heat added to the reaction). This is the exact opposite for gases in a liquid solution as the solubility will generally decrease with increased energy (heat).

F. Pressure will have very little to no effect on the behavior of liquids in a liquid solution. The same is generally the rule with solids as well. Gasses on the other hand will increase solubility with increasing pressure.

G. Osmosis is the diffusion of the solvent through a semi permeable membrane into a more concentrated solution. Solutions with the same osmotic pressure are called isotonic solutions. The osmotic pressure is defined mathematically by: $\pi = CRT$. π = osmotic pressure, C = concentration, R = gas constant and T = Temp Kelvin.

XII. Liquids. The definition of a liquid is a substance that the molecules are in constant random motion. Liquids will also have a definite volume but they will assume the shape of the container. In liquid form, there is little effect of pressure on the volume that they occupy and temperature has only slight effects on the volume that a liquid will occupy.

A. Liquids demonstrate the property of surface tension. This is the strength of the inward force of the molecules of the liquid. His value will decrease as the temperature of the liquid increases.

B. Evaporation is when the molecule of the liquid obtains sufficient kinetic energy to overcome the attractive forces of the other molecules. For this to happen the molecule must reside close to the surface of the liquid and the environment. If the molecule is able to overcome the attractive forces this is then called a phase change of evaporation. There is a specific term for this process that is the heat of vaporization. It is expressed as follows: the number of calories (remember calorie is a measurement of heat energy) required to convert 1 g of liquid to 1 g of vapor without changing the temperature.

C. Boiling point of a liquid is defined as the temperature that the pressure of the vapor escaping from the liquid is equal to the atmospheric pressure (760 mm Hg)

Cell Structure

I. Organism types

A. Prokaryotes do not have a nucleus. Archebacteria and eubacteria are the residents of this classification.

i. No Organelles are present

ii. Single chromosome

iii. Reproduce by binary division

iv. Utilize anaerobic or anaerobic metabolic pathways.

v. Have cytoskeleton to support the cell wall present.

vi. There is a rigid cell wall that will surround the plasma membrane.

vii. There exists a periplasmic space that is found between the plasma membrane and the cell wall. The function of the periplasmic space is as a potential space for the secretion of proteins to accumulate.

viii. Plasma membrane is a selective barrier for the entry of specific compounds. It is made of phospholipids and it is the site of the electron transport chain.

B. Eukaryotes are cells that contain a nucleus. This is the most variable categorization of organisms. This includes the plants, animals, molds, fungi, and ameba.

i. Have defined nucleus and nucleolus

ii. Multiple chromosomes

iii. Reproduction is via mitosis or meiosis
iv. Aerobic metabolism
v. Extracellular matrix with a supportive cytoskeleton.
vi. There are many organelles within the matrix of the cell.

C. Viruses are parasites. There is a classification of these that are called bacteriophages as they originate from bacteria.

II. Membranes of the Eukaryotic Organism

A. The membrane is designed to act as a barrier, not just to keep things out but also to keep things within the cell. It also acts as a receptor site and recognition site.

B. Structural characteristics demonstrate a large proportion of proteins, lipids and carbohydrates. The ratios are variable and it is this property that allows for the fluidity of the membrane itself. The composition of the membrane is reflective of the diet, function and the environment of the cell.

i. membrane lipids are phospholipids, glycolipids, cholesterol and these are amphipathic molecules. These membrane lipids will spontaneously form lipid bilayers in aqueous solutions. This property will have the hydrophilic side (the polar head) facing the exterior and the hydrophobic (nonpolar tail) facing the inside.
ii. The lipid content allows a special property of the membrane, the ability to laterally diffuse, or flex, or flip flop.

iii. Increased cholesterol levels will negate the flexibility of the cell membrane.

iv. The distribution of the phospholipids within the cell membrane is relative asymmetric with choline phospholipids facing the outer face of the bilayer and the amino phospholipids facing the inner or the cytoplasmic side of the membrane. Glycolipids are also found to reside in the outer facing side of the membrane. These glycolipids are important for cell to cell signaling.

Fig 3

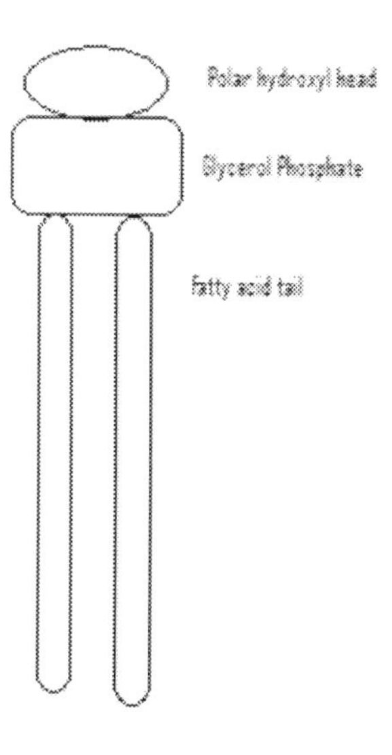

C. Membrane proteins are either integral membranes or peripheral proteins. Integral proteins are difficult to dissociate from the membrane while there are some integral proteins that are exposed on both sides of the membrane. Integral Proteins dissolve in lipid bilayer of the plasma membrane (PM). Transmembrane proteins span the entire PM, function as receptors

& transport proteins. Most transmembrane proteins are glycoproteins, amphipathic, with hydrophilic and hydrophobic amino acids. Most are long, folded polypeptides that pass back & forth across the PM, also known as 'multipass' proteins. Most transmembrane proteins are anchored to inner leaflet of PM via fatty acyl or prenyl groups, all are globular proteins. In freeze-fracture, most integral proteins remain bound to P face (the external surface of the inner leaflet)

The proteins that are exposed on both sides are called transmembrane proteins. This is in comparison to the proteins that are found to reside on the exterior of the polar heads that are known as peripheral proteins. These proteins are attached via electrostatic bonds or by hydrogen bonds. Primary Functions of the membrane include: maintaining structural & functional integrity of both the cell and of its organelles, acts as semipermeable membrane between the cytoplasm and the external environment, permits cell to recognize and to be recognized by other cells and macromolecules, and tranduces extracellular signals into intracellular events through several mechanisms. Peripheral Proteins ***do not*** extend into the lipid bilayer. They are located on cytoplasmic aspect of inner leaflet. Outer leaflets of some cells possess covalently linked glycolipids to which peripheral proteins are also anchored and extend into the extracellular space. Peripheral proteins bond to phospholipid bilayer polar groups, or to integral proteins via non-covalent interactions. Peripheral proteins usually function as part of cytoskeleton, or as part of an intracellular 2nd messenger system (ex: Adenylate cyclase). Include a group of anionic, Ca+2 dependent, lipid binding proteins known as annexins, which can act to modify the relationships of other peripheral proteins with the lipid bilayer.

D. Fluidity of the membrane is regulated to a large extent by the nature of the packing and the interaction of the fatty acyl chains of the phospholipids.

i. long chain fatty acids that are saturated will pack densely. This will increase the interactions and thus the reduction of the fluidity of the wall.

ii. Increased temperature will alter the shape of the C—C bonds by forcing them into a gauche position. This will weaken the bond interaction and therefore make the membrane more fluid.

iii. The longer the chains of the tails will Tm (melting point) will increase and make the membrane more rigid.

iv. Unsaturated fatty acids with the "cis" bond shapes will not pack densely and this property will reduce the rigidity. This unsaturated bond angle will also reduce the melting point.

v. Cholesterol content will affect the fluidity as well, the more present, the more rigid the membrane.

vi. Membrane protein content varies from about 20% to 80%, according to the function of the membrane. The average Protein content 50% for most. Proteins serve many functions in ion and nutrient transport, signal transduction, receptor structures, and others.

Many proteins span the lipid, anchored via ionic, electrostatic, and hydrophobic bonds.

vii. Lipid to protein ratio ranges from 1:1 (most cells) to 4:1 (myelin). Some membrane proteins diffuse laterally in lipid bilayer, others are immobile. Some held in place via cytoskeletal components

vii. No 'flip-flop' movements of membrane proteins or phospholipids due to polarity, non-covalent ligands. Lateral movements of membrane lipids common, fluidity contributes to receptor functions

III. Organelles of the Eukaryotic cell

A. Nucleus is the control center of the cell. It is bound by a double layered membrane that is penetrated by opening called nuclear pores. Each of the pores has a specific structure of eight multiprotein granules creating the opening. The nuclear membrane is lined by the nuclear lamina that is a fibrous network of proteins called lamins. The laminas are responsible for the breakdown of the nuclear envelope with mitosis.

B. Nucleolus is the site of the synthesis of ribosomal RNA.

i. Ribosomal synthesis is also initiated in the nucleolus.

ii. This is located within the nucleus, therefore it is a suborganelle.

iii. Its size is definitive of the cellular activity.

C. Endoplasmic reticulum (ER) is pivotal in the synthesis of the proteins, lipids, carbohydrates and steroids of the cell.

i. This is the initial site of the assembly of the proteins synthesized and direction of the tertiary and quarternary structural configuration.

ii. The N and the O linked oligosaccharides are attached in the ER.

iii. This is continuous with the outer nuclear membrane.

iv. There are 2 subtypes of the ER, the smooth and the rough ER.

a. Rough ER is the site of protein synthesis and as such it is prominent in those cells that synthesize proteins. The description of he rough is due to the attachment of the ribosomes to the outer aspect of the membrane.

b. Smooth ER is the site of lipid synthesis. It is also the site of tissue detoxification as well with the cytochrome P450 system. It also synthesizes steroid hormones.

v. The smooth ER is a continuation of the rough ER.

vi. It is very dynamic and able to change structure as the cellular requirements change.

D. Golgi Complex is the traffic cop of the cell as it directs the intracellular macromolecular movements.

i. Both the N and the O linked oligosaccharides are modified here.

ii. Lysosomal destined proteins receive the appropriate identification tag here in the form of the phosphate added to the mannose of the N linked oligosaccharides.

iii. Has two distinct faces, the cis and the trans face.

iv. Looks like a flattened stack of tubing.

v. Budding of the vesicles is indicative of the proteins as they are being sent out to the appropriate destination.

vi. Cis- face called cis-Golgi network (CGN). Medial compartment consists of a few cisternae placed between cis- and trans-faces.

vii. Trans Golgi network (TGN) lies apart from the last cisterna, sorts proteins for final destination.

viii. Golgi system processes membrane packaged proteins synthesized in RER, and recycles & redistributes membranes.

viv. ERGIC – [Endoplasmic Reticulum and the Golgi Intermediate Complex], functions as first way station for segregation of anterograde vs. retrograde transport, concentrates proteins.

x. Vesicles are directional; migrate toward appropriate membrane for correct fusion

xi. Clathrin coated vesicles

Vesicles coated with clarthrin, a polypeptide that forms a triskelion (3 legged structure); 36 clathrin triskelions associate to form a cage-like lattice that surrounds the vesicle, facilitated by adaptins that perform a recognition role for interface between pit and cargo molecule

xii. Coatomer coated vesicles. Coatomer is a large protein that depends on an ADP-ribolylation factor(ARF), which binds GTP and recruits coatomer subunits & selects cargo molecules. The coatomer coated vessicle is designed to transport proteins from RER to ERGIC, Golgi to Golgi, mediate bulk flow or constituitive pathway; cargo is bound via v-snares, which also guide movements of captured vesicles.

xii. Caveolin coated vesicles. Invaginations of the PM of endothelial cells and smooth muscle cells associated with cell signaling functions, transcytosis, & endocytosis. Least common of vesicle types; caveolin is also a protein.

E. Lysosomes are the digestion mechanisms of the cell. They contain acid hydrolases. These organelles are very acidic with a pH of around 5. These organelles are actually membrane bound sacs of the acid hydrolases. There are close to 22 different acid hydrolases.

i. 4 types named after content of vesicular material: multivesicular, phagolysosomes, residual bodies, and autophagolysosomes.

ii. Dense, membrane-bound organelles, have diverse size, shape, identified by acidic nature (pH ~ w5), contain acid phosphatase plus about 50 acid hydrolase enzymes.

iii. Lysosomes are formed when sequestered material fuses with a late endosome, and enzymatic degradation begins.
iv. Average diameter about 0.5-1 µm .
v. Multivesicular lysosomes. Multivesicular bodies are formed by fusion of an EE with a LE
vi. Phagolysosomes are formed by fusion of a phagocytic vacuole with a late endosome or a lysosome
vii. Autophagolysosomes are formed by condensation of an autophagic vesicle with a late endosome or lysosome; function in cellular degradation of old cells
viii. Residual bodies are any type of lysosome that has expended their capacity to degrade material.

F. Peroxisomes will degrade the amino acids and the fatty acids through the action of enzymes like catalase. The byproduct of the reactions will be the production of hydrogen peroxide. These are structurally very analogous to lysosomes. Also known as microbodies are small (0.15-0.50 µm dia) membrane bound spherical, or ovoid granules typically identified via catalase reaction. The perioxisome may contain a nucleioid (core of urate oxidase or uricase) [no nucleoid in human cells]. These peroxisomes are known to originate from pre-existing peroxisomes, recognized by receptor proteins (peroxins) and will divide by fission, $t^{1/2}$ = 5-6 days, rep 1-2% hepatocyte volume. Core is Crystalline Catalase & Proteins with a single Membrane Surrounding. All require O2 to perform oxidization. The function of peroxisomes is oxidation of LCFA, cholesterol synthesis, detoxification (ETOH & other organic chemicals. Early Endosomes (EE). Irregular peripherally located vesicles contain receptor-ligand complexes known as CURLS (Compartment for Uncoupling of Receptors and Ligands). Acidic nature maintained by ATP proton pumps (pH ≅ 6). Acidity assists in uncoupling of receptors from ligand. Late Endosomes (LE) LE plays a key role in lysosomal pH, also known as intermediate compartment, where the pH is ≅ 5.5, lie deep in cell. They receive ligands via microtubular transport of early endosomes. Contain both lysosomal hydrolyases and lysosomal membrane proteins (from RER), become lysosomes upon activation.

G. Ribosomes are the actual site of protein synthesis.

H. Mitochondria are the energy producers of the cell.

i. they have a double membrane with very specific activities on each membrane. Power house of cell, rod shaped (0.2 x 7 microns long), contain all enzymes of TCA cycle and electron transport

ii. the inner mitochondrial membrane is the site of the electron transport chain. Granules within matrix bind Ca2+, Mg2+

iii. the mitochondrial matrix is the location of the TCA or Krebs cycle or the citric acid cycle.

iv. the inner membrane has many folds within it and they are there to increase the surface area and the term for them is the cristae.

v. The mitochondria are maternally inherited

vi. mitochondrial RNA is found within the matrix. Mitochondria contain circular DNA, approx 1% of total DNA of cell, replicate independently of cell, life span ~10d. ATP formed via chemiosmotic coupling mechanism, in cristae on inner membrane. Also form ATP from oxidation of FA, glucose, and amino acids.

IV. Cytoskeletal structures of the Eukaryotic cell

A. Microfilaments, intermediate filaments and microtubules make up the cytoskeletal substructures.

i. Microfilaments are small only 7nm wide. The major protein that they are composed of is actin so often they are called actin filaments.

ii. Myosin is the major interactor with actin as it associates to elicit contractile processes.

B. Intermediate filaments are long, rod like structures.

i. intermediate filaments are able to self assemble.

ii. the main function of the intermediate filament is in structural support.

iii. there are 5 main classifications of intermediate filaments

a. vimentin – seen in fibroblasts and epithelial cells

b. desmin – seen in muscle cells

c. neurofilaments - seen in axons

d. Glial fibrillary acidic protein – surrounds neurons within the glial cell.

e. Cytokeratins – there are 30 subtypes of these proteins.

C. Microtubules consist of tubulin.

i. tubulin has 2 subtypes α and β.

ii. structurally, they are very unique as they are composed of 13 parallel aligned protofilaments.

iii. there are many associated proteins that function with the microtubules called microtubule associated proteins (MAP's)

iv. microtubules are responsible for both intracellular and extracellular movement

v. they are also responsible for the formation of the mitotic spindles

vi. when associated with kinesin, there is a produced minus to plus directed movement of the vessicles within the microtubules.

vii. MAP1C is the protein that opposes the movement of kinesin and moves vesicles from the plus to minus direction.

Fig 4

Organelle	Function
Plasma Membrane	Transport of ions and molecules Recognition Receptors for small and large molecules Cell morphology and movement
Nucleus	DNA synthesis RNA synthesis
Nucleolus	RNA synthesis and ribosome synthesis

Endoplasmic Reticulum	Membrane synthesis Synthesis of proteins and lipids for export Export proteins Detoxification reactions
Golgi Apparatus	Protein exportation Post-translational protein modification
Mitochondria	Cellular respiration Synthesis of Urea and heme Oxidation of lipids and proteins Energy Production
Microtubules / Microfilaments	Cellular motility and mobility Cytokeletal structure
Lysosome	Cellular debris clearing Hydrolysis of lipids, proteins, carbohydrates and nucleic acids
Perioxomes	Oxidative reactions of cellular byproducts
Cytosol	Metabolic reactions involving the carbohydrates, amino acids, nucleotides while synthesizing proteins

I. Water is a very important material to living organisms. In many biologic systems, water is a reactant as well as chemical determinant. The chemical properties of water are such that is able to possess both a negative and a positive charge. This property is called a dipole moment. The negative portion of the molecule is near the oxygen atom and the positive portion of the atom is near the hydrogen. The overall charge that this molecule displays will be a net charge of zero.

Fig 5

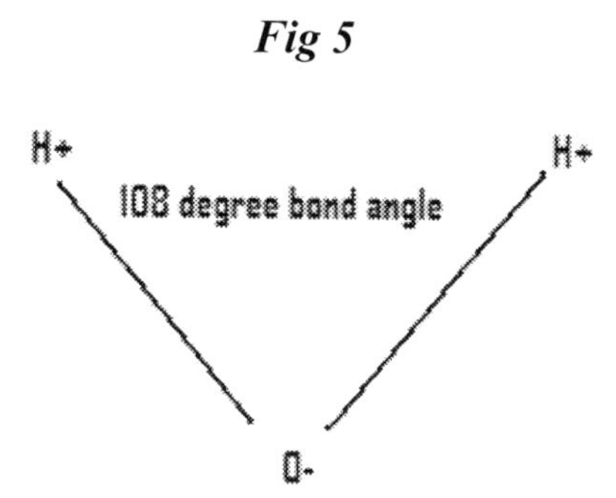

A. The interaction between the hydrogen and the oxygen atoms will produce a hydrogen bond. These are electrostatic interactions that develop between the hydrogens and the development of the attraction from the unpaired electrons of the oxygen atom.

B. The hydrogen bonds will develop between the water molecules or with any molecule that water may come in close proximity with.

Amino Acids & Proteins

1. Structure – proteins are long unbranched polymers derived from the 20 amino acids.
2. Function of Proteins
   a. Enzymes – most enzymes are proteins
   b. Transport & storage of smaller molecules and ions.
   c. Structural – strength and structure – component for the intracellular and extracellular movement
   d. Structure of Skin & Bone – Collagen, most abundant protein in body.
   e. Immunity
   f. Hormonal regulation – as cell surface receptors
   g. Control genetic expression
3. Disease states are often due to altered function or structure of proteins. (hemaglobinopathies)
   a. Marfan's Syndrome – single amino acid change – in the elastic tissue protein fibrillin
   b. Cystic Fibrosis- single amino acid deletion in the ATP binding domain of transmembrane conductance regulatory protein

II. Amino Acids – functional unit of proteins

A. Have amino group ($-NH_3$) carboxyl group (-COOH) hydrogen atom, and (R) side chain that are all bound to the central alpha carbon. The R side chains are attachment sites for other compounds. The common amino acids are those known to consist of at least a single codon. Other amino acids are able to be formed through enzymatic modification of a common amino acid. These amino acids are were described in the previous chapter, cystine, desmosine, isodesmosine, hydroxyproline, hydroxylysine, phosphoytyrosine, phosphoserine and gamma carboxyglutamate.

Fig. 6

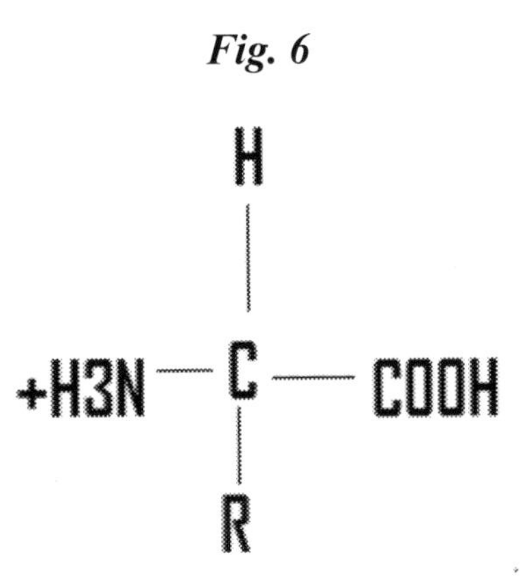

1. The amino acids Serine, Threonine and Tyrosine all contain polar hydroxyl groups and therefore, able to participate in hydrogen bonding. Additionally, serine, threonine is able to bind phosphate groups through the hydroxyl groups. This is not the case for tyrosine.
2. Asparginine and glutamine each have a definitive carbonyl group and an amide group. Each is able to participate in the hydrogen bonding. The amide of the asparagine will serve as an attachment for oligosaccharides in the synthesis of glycoproteins.
3. Disulfide bonds are in proximity to allow the interaction of the two cystine residue that will yield the covalent cross linkage of the disulfide bond. The

base of this reaction is the sulfydryl group from the Cysteine.

B. Exception is proline, it is imino acid. The structure of this amino acid is that is has the α amine as a secondary amine and the α nitrogen allowing for 2 covalent bonds to the carbon. This ability creates limitations to the incorporation of this amino acid into polypeptides.

C. Post-translational modification – these modifications are made to direct specific locations and functions.

D. Optical activity of amino acids – there are L- amino acids found in the human body and there are D – amino acids that are found in bacterial and antibiotics

E. Properties of the amino acids are defined by the side chain that is attached to the alpha carbon. This carbons is also considered as a tetrahedral carbon, the exception being glycine.

1. Non polar side chains are: Glycine (GLY), Alanine (ALA), Valine (VAL), Leucine (LEU), and Isoleucine (ILE)
2. Aromatic Side Chains are: Phenylalanine (PHE), Tyrosine (TYR), and Tryptophan (TRY)
3. Hydroxyl containing amino acids are: Serine (SER), Threonine (THR)
4. Acidic side chains are: Aspartate (ASP), Glutamate (GLU)
5. Amidic amino acids are: Asparginine (ASN), Glutamine (GLN)
6. Basic side chain amino acids are: Lysine (LYS), Arginine (ARG), Histidine (HIS)

7. Sulfur containing side chains are: Cysteine (CYS), Methionine (MET)
8. Imino acid side chain is: Proline (PRO)

F. Chemical Structure of the 20 amino acids:

Fig 7

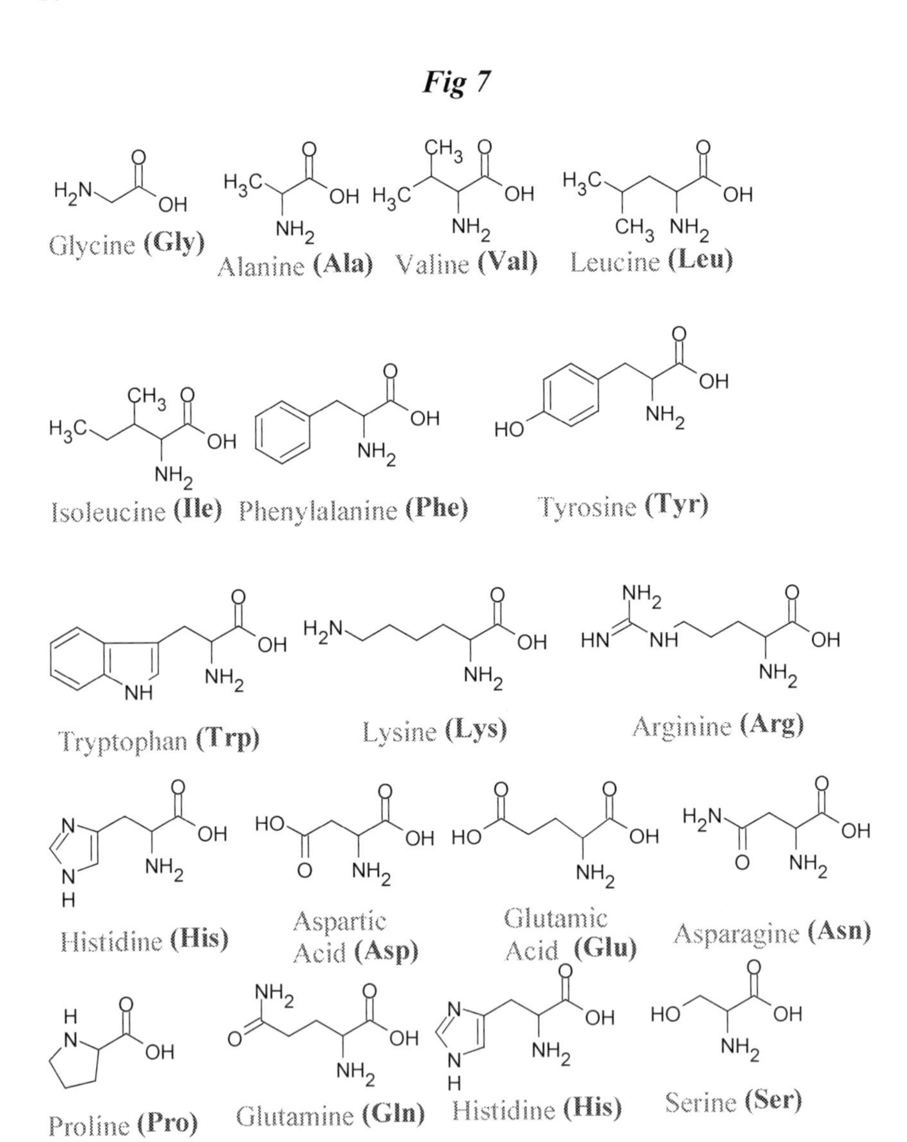

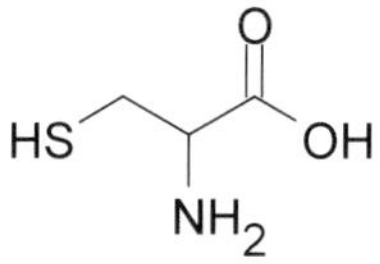

Cysteine **(Cys)** 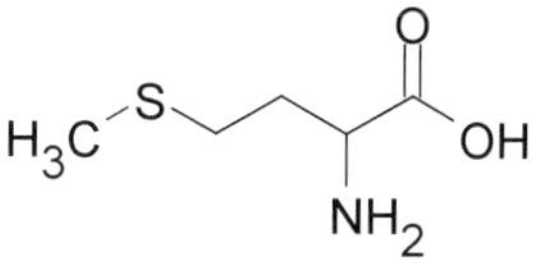   Methionine **(Met)**

III. Amphoteric properties

A. These are both acidic and basic

B. Dipolar (zwitterions) molecules –have both positive and negative charges. It I necessary to calculate the pH where this zwitterion form is found. This specific point is the isoelectric point (pI). This value is constant at a for the compound at a particular ionic strength and temperature. At a pH less than the pI, the overall charge will be positive, while the converse is found at a pI greater then the pH.

a. The alpha carboxyl is negatively charged

b. The alpha amino group is positively charged.

c. This makes the overall charge on the molecule neutral

C. At low pH (high concentrations of hydrogen ions) the carboxyl group accepts the proton

D. At high pH (low concentrations of hydrogen ions) the amino group loses its proton.

E. Amino acids with dissociating groups

1. Amino acids - aspartate, glutamate are acidic while lysine, histidine, arginine are basic.

2. Cysteine & tyrosine have negative charge on the side chain when dissociated.

F. The acid – base properties of amino acids demonstrate properties of a weak acid and a weak base in aqueous solution.

1. The derivation of the dissociation of the proton released is from the Henderson- Hasselbalch equation that states HA (weak acid) ↔ H+ (proton) + A^-(salt form or conjugate base)
2. The equation is then expressed as $K_a = [H+] [A^-] / [HA]$
3. The larger the K_a, the stronger the acid.
4. The buffer aspect of the amino acid is demonstrated by the manner that the solution resists change in pH following the addition of a base or acid.
5. The pH is equal to the pK_a when the [HA] and [A-] are equal.
6. The isoelectric point (pI) is the pH where the amino acid is neutral. As mentioned previously, a pH less than the pI, the overall charge will be positive, while the converse is found at a pI greater then the pH.
7. The dissociation of an acid is represented by the dissociation constant (K'_a) and the pK'_a.

IV. Peptides and Polypeptides – written from left to right with the amino acid containing the free alpha amino group (the N- Terminus) first to the amino acid containing the free carboxyl (the C- terminus). The formation of the peptide is generally a dehydration reaction through the bonding of α amino group of one amino acid to the α carboxyl group and eliminated a molecule of water. The peptide bond will form in either the trans, most stable configuration or the cis configuration.

A. Peptide bond – between the alpha carboxyl of one amino acid to the alpha amino group of another amino acid.

B. This is formed from the removal of water as mentioned above.

1. This is very energy expensive (endergonic) and requires the hydrolysis of high energy phosphate bond.
2. Possible for these bonds to be trans or cis. Trans is the favored bond angulation as it is the most stable. The cis configuration is going to allow some repulsive forces to become apparent, through the manner that the 2 side chains were bound on the same side of the peptide bond.

C. Polypeptide chains – it is the result of the linkage of many amino acid polypeptides. These chains are normally over 100 residues long. Exceptions are available, (vasopressin is 9 residues, glucagons is 29 residues, thyrotropin releasing hormone is 3 residues) The protein titin is a protein that consists of more than 27,000 amino acid residues.

D. Amphoteric properties - the formation of the peptide bond removes two dissociation groups, 1 – alpha amino and 1 alpha carboxyl per residue.

1. The nature of the proteins ability to fold or to bond is influenced by the hydrophobic or the hydrophilic nature of the side chains.

2. When there is clusters of nonpolar side chains are found on the surface of the protein, it is associated with a specific function as a receptor or a tag.
3. When the charged side chains are found on the inside of the membrane, it is assumed that they function in a role of stability of the cell.
4. Transmembrane proteins demonstrate the hydrophobic side chains on the outside and the ionic groups on the inside. This structural configuration allows for the formation of channels.

V. Purification of Proteins

1. Protein solubility is influenced by salt concentration of the solution.
   a. Salting out is the addition of divalent salts (ammonium sulfate) to a solution of protein. There will be precipitation of proteins at variable salt concentrations. This method is used to increase the amount of a specific protein in a fraction.
   b. Salting in is the addition of inorganic ions to allow dialysis against a solution of a low sat concentration and produce precipitation of specific proteins from the solution.
2. Separation by molecular size.
   a. Dialysis – uses a predetermined semipermeable membrane that the solution is passed through to select the size molecule. Most dialysis membranes will restrict the flow of molecules greater than 15Kdal.

b. Gel filtration – (molecular exclusion chromatography, molecular sieving) uses a column of insoluble but highly hydrated carbohydrate polymer in the form of porous beads. Small molecules can pass through the pores but the larger ones cannot. The rate at which a molecule flows through the column is dependent on the size and shape of the molecule. This is used to estimate the molecular weight of a protein as well as separate a protein.

c. Ultracentrifugation – data is in the form of Svedburg Units (S). These rates are derived from the sedimentation constant (s) with s x 10^{13}

d. Sodium dodecyl (SDS) polyacrylamide gel electrophoresis – this is done on a cross linked polyacrylamide gel with SDS and a specific reducing agent (beta mercaptoethanol) separates proteins on the basis of molecular weight. SDS dissociates the quarternary structure into monomers. SDS forms negatively charged micelles with the protein, the effects of the protein charge is lost and therefore the shape of all the proteins becomes linear. These micelles are also separated for size by migration using gel electrophoresis. The SDS protein micelles migrate to the positive pole. Smaller molecules move faster.

3. Separation on the basis of molecular charge

a. Ion- exchange chromatography – the proteins are bound to the ion exchange resin in the column. The tightness of the bond is due to the number of residues available for binding with the ion exchange resins. There is a column of insoluble ion exchange material that carries either polyaionic or polyanionic groups. These groups bind oppositely charged groups by ionic concentration.

b. High performance liquid chromatography (HPLC) similar to the ion exchange but with high pressures (5,000 – 10,000 psi). Faster and more accurate.

c. Electrophoresis – uses an electric field to drive the movement of a molecule that has a net charge. In electric field the protons migrate by the net charge that is determined by the nature of the ionizing groups and the pH. For every protein there is a pH termed the isoelectric point (pI) where the molecule has no charge. The migration is determined by the electrophoretic mobility (μ). This is a ratio of the velocity of migration (V) to the electric field strength (E). $\mu = V/E$ and measured in cm^2 per volt second.

    i. Gel electrophoresis – used to separate plasma proteins for diagnostic purposes. Sample is layered on a matrix of polyacrylamide gel or agarose gel. After the electrophoresis is completed the proteins are stained with differing proteins having differing zones (bands) dependent on the overall charge of the protein.

    ii. Free capillary zone electrophoresis – is using electrophoretic separation by s free solution that is placed in container of small bore capillary tubes (0.05 – 0.03 mm). There is no matrix required in this methodology. Since there is efficient dissipation of the heat this system uses very high voltages (10kV). This method works very quickly and is used to separate charged molecules.

1. Separation by specific affinity binding
   a. Affinity (absorption) chromatography – uses the fact that specific proteins have the ability to bind to another molecule (ligand) very strongly. This is a noncovalent bond. The ligand can be attached to a large hydrated particle with the solution of the protein in question being passed through the column. The protein would bind strongly to the ligand. Once washed the beads containing the ligand bound protein are then eluted by a solution of the pure ligand.

VI. Conformation of Proteins. Each protein has a native state and a unique 3 D structure that is called its conformation.

A. Primary Structure – it is the backbone of the polypeptide. It is the specific amino acid sequence.

1. Peptide bonds are the bonds responsible for the linkage of the α carbon on one amino acid to the α carboxyl.
2. Peptide bonds are very strong bonds they are not affected by the things that normally denature protein bonds. It requires strong acids or bases or extremely elevated temperatures to hydrolyze these bonds.

i. The peptide bond is unique in that it displays characteristics of a double bond, shorter, and therefore more rigid and planar.

ii. The peptide bond presents in a trans configuration due to the increased steric hindrance created by the R side chains.

B. Secondary Structure – it is the spatial description. This structural description is totally dependent upon the primary structure. It comes from the interaction of the neighboring amino acids.

1. Hydrogen bonding – is a characteristic of the secondary structure. It is due to the formation of the hydrogen (-H) bonds between the - CO group of one peptide and the –NH of another close peptide. If these bonds are formed in the same chain then the result is either an alpha helix or beta turns. Proline is an amino acid that is disruptive to the formation of the structure due to the steric hindrance of the side chain. If the concentration of charged amino acids is large there will also be disruption of the bonds as there will be a tendency for the charged side chains to form ionic bonds.

i. Alpha Helix – each –CO is hydrogen bound to the –NH of a peptide bond 4 residues ahead on the same chain. This produces 3.6 amino acids per turn. The rigidity of this structure is accomplished through the number of disulfide bonds within the helix.

ii. Beta pleated sheets – like those in silk. These are pleated due to the C-C bond being tetrahedral and therefore cannot be planar. The chains are side by side with the hydrogen bonding between the –CO of one peptide and the –NH group of another peptide on the neighboring chain. These chains may run in the same direction or in opposite directions. These are called parallel Beta pleated sheets and antiparallel Beta pleated sheets.

a. The structure of this molecule is that there are 2 or more polypeptide chains, the Beta strands. The bonding in these strands is specific to the manner that the hydrogen bonds are formed, in this Beta sheet, the hydrogen bonds are perpendicular to the polypeptide backbones.
b. There is a special arrangement of these Beta sheets that will allow them to form "parallel" or "antiparallel" configurations.
c. In the parallel configuration, all the N terminal will be in the same directional position so that the position is consistent, either C to N or N to C, but always the same.
d. In the antiparallel position, the arrangement is such that there is a C to N terminus that is aligned to the N to C terminus.
e. It is possible in this configuration to produce inter chain hydrogen bonding.
f. It is also possible for the configuration to be obtained via a single polypeptide chain

as it folds over itself. In this configuration, the hydrogen bonds are intra chain bonds.

g. In globular proteins, the Beta sheet is generally seen to have a right handed twist as looked upon along the long axis of the chain.

2. Beta turn AKA Reverse turn – it is the tightest turn seen, possible to result in a complete reversal of direction. There is a large concentration of glycine but with a pattern of four amino acids, one of them being proline.

3. Supersecondary structures are developed by the closely packed side chains that are available from adjacent amino acid secondary structures.

C. Tertiary Structure – it is the spatial relationship of the more distant residues. It represents the lowest energy state, and thus the greatest stability. This configuration is from the hydrogen bonding within the same chain or between chains. Again it is the primary structure of the amino acid that will be the dictating factor for the development of the tertiary structure.

1. The tertiary structure also has differing domains. Domains are the functional 3D presentation of the amino acid. Chains that are greater than 200 amino acids long are going to present with two or more domains.

2. The structure is stabilized by the formation of the interchain sulfide bonds, hydrophobic interactions, hydrogen bonds and ionic interactions.

a. Sulfide bonds are from the –SH groups

b. Hydrophobic interactions are from the internally arranged hydrophobic domains of the nonpolar amino acids. These internally arranged side chains will create hydrophobic interactions that show inverse relationship to the distance of the side chains in question. The closer the side chains, the greater the attraction.

c. Hydrogen bonds are formed from the side chains of the oxygen and nitrogen bound hydrogen exposed atoms. This is very apparent in serine or Threonine. The oxygen will form strong bonds with the carbonyl or carboxyl groups to the opposing peptide.

d. Ionic interactions are from the negatively charged amino acids of glutamate and aspartate as they bind as an ionic bond to the positively charged groups.

D. Quarternary Structure – refers to the spatial relationship between actual polypeptide chains.

1. These quarternary structures refer to two or more monomeric proteins. These are generally held together by noncovalent interactions like those of the hydrophobic interactions, hydrogen bonds or ionic bonds.

VII. Fibrous proteins like collagen, keratin, elastin have been the brunt of the fibrous proteins that have been studied.

A. Collagen is the most abundant protein in the human body. It has a variety of configurations within the tissue as well, from gelatinous to very stiff and rigid. Collagen precursors are firmed within the fibroblasts, osteoblasts or the chondroblasts. The precursors are then secreted into the matrix for further modifications. As the development is completed, the collagen molecules will cross link to create structure and rigidity.

B. Structure of a collagen molecule is seen to be a tripolypeptide with each strand of the 3 being an α chain. The three chains are bonded by interchain hydrogen bonds.

C. Specific amino acid sequence will be with glycine in every third position. With the next position most likely hydroxyproline and the next position being hydroxylysine.

  i. There are a variety of different types of collagen available, the most common is type 1.
  ii. Type 1 is derived from two α1 chains and one α 2 chain.

D. The hydroxylysine and hydroxyproline are not seen in the majority of other proteins.

  i. These subunits undergo hydroxylation but this reaction is run only after they are incorporated into the peptide chain. This type of cellular modification is called posttranslational modification.
  ii. Hydroxyproline is the most important of the two in the structure of collagen.
  iii. Additional modification can come from glycosylation using galactose or glucose.

E. Synthesis of collagen begins with a nascent peptide chain that contains collagen molecules. These collagen molecules are special in that they contain a special sequence at the N terminus. The function of this signal sequence is to tell the cell the protein is destined for export to the matrix.

i. This sequence is rapidly cleaved once the polypeptide enters the endoplasmic reticulum.
ii. The product of this cleavage will be a pro α chain.
iii. Pro α chains will be modified by the action of enzymes within the endoplasmic reticulum. This modification is done as the synthesis is continued.
iv. The hydroxylation reactions referred to earlier will require molecular oxygen. Additional requirements are the need for a reducing agent like ascorbic acid (Vitamin C)
v. The reaction will produce prolyl hydroxylase and lysyl hydroxylase but will require the ascorbic acid to function. The purpose of this reaction is to begin the crosslinks that are providing the strength to the collagen fibers.
vi. Glycosylation of a variety of the hydroxylysine residues with a glucosyl-galactose or glucose.
vii. Upon the completion of the previous steps, the collagen molecule is still developing and it is now called a procollagen molecule. This particular molecule has a distance fibrillary arrangement with a central triple helix surrounded by non-helical amino and carboxyl propeptides.

viii. The next step is the translocation of the molecule to the golgi complex for packing into the secretory vesicle that will exocytose the procollagen molecule into the extracellular space.

viv. The releasing signal is the cleavage of the N or C procollagen peptidases that will be cleaved to indicate the molecule is ready for release.

x. The collagen fibrils are unusual in the method that they form. They actually spontaneously aggregate. This spontaneous aggregation will however, produce very orderly array of fibrils that will produce a very strong shape within the molecule. The increased strength is from the way that the fibers will overlap the adjacent fiber by ¾ of the length of the adjacent fiber.

xi. Lysyl oxidase is the enzyme that is the byproduct of the joining of these fibers. It functions in the oxidative deamination of some of the lysyl and the hydroxylysyl residues. This reaction will generate a second byproduct, and aldehyde that will form the crosslinks with the neighboring collagen molecule.

F. Elastin is the rubber like connective tissue. It is found throughout the body, but the lungs, the blood vessels and the elastic ligaments being the main distribution.

i. The composition of the amino acids in elastin is mainly smaller nonpolar amino acids like glycine, Alanine and valine. Additional amino acids, proline and lysine are seen in great concentrations.

ii. In elastin, there is little to no hydroxyproline and no hydroxylysine.

iii. The crosslinks that are formed for elastin are from lysine. These crosslinks are not like the ones formed in collage, as they are not in an orderly array, but rather no apparent order is seen. There is a cloverleaf like arrangement of lysine residues or elastin molecules that will be covalently linked to produce what is called a desmosine cross link. It is this formation that allows for the rubbery like properties of elastin.

G. Keratins are proteins that will form tough fibers. The keratins are seen mostly in the hair, nails, outer epidermal skin layers. They produce many cross links that are the rationale for the strength of these structures.

VIII. Analysis of Protein Configuration

A. Primary structure – determining the sequence can be done using acid hydrolysis (110 degrees C in 6N HCL in sealed tube x 24 hours). Also able to use Ion exchange chromatography in cation exchange columns made from sulfonated polystyrene with ninhydrin reactions. This reaction turns blue with alpha amino acids (proline). Also can use N terminal residue – this can be done using a reagent that bonds to the alpha –NH like (FDNB) Fluorodnitrobenzene or dinitrofluorobenzene (DNFB) or Sangers reagent. Polypeptides that have been reacted with this reagent show aromatic substation that result in the hydrolysis from the polypeptide. Possible to use Dansyl chloride as a reagent that will form fluorescent

N terminal derivatives that can be detected at very low concentration.

B. Identifying the Amino acid sequence

a. Edmans reaction – uses phenylisothiocyanate to react with the N terminal group that will form a phenylthiocarbymal derivative. This reaction is carried out under slightly alkalitic conditions.

i. The reaction yields instability in the N terminal peptide bond.

b. Sequencing of the peptide fragments –

i. Trypsin – cleaves carboxyl side of basic amino acids (lys, arginine)

ii. Chemyotrypsin – carboxyl side of aromatic side chains (phenylalanine, tyrosine, tryptophan)

iii. Staphylococcal protease – carboxyl side of acidic amino acids (aspartate, glutamate)

iv. Chemical Agents – Cyanogen bromide – carboxyl side of methionine. Hydroxylamine – cleaves asparginine – Glycine bonds.

v. Acid hydrolysis of the solution will yield the release of the individual amino acid with the additional benefit of hydrolyzing glutamine and asparginine to glutamate and aspartate. Ninhydrin will then be reacted to the solution to label the amino acid side chains with amino acid, ammonia, and amines by turning them purple. This is in contrast to the imino nitrogenous base that is seen in proline as it is turned yellow.

c. Locating the Disulfide bonds – found between cycteinyl residues and form crosslinks. Performic acid oxidizes these residues to cystic acid residues.

C. Secondary and Tertiary structures

d. X- Ray crystallography – the protein must be crystallized for the structure to have elongated repeating structure as in fibrous proteins.

e. Studies of Proteins in Solution –

i. UV spectroscopy – proteins absorb UV light as a result of the absorption of the light by the peptide bonds and the tyrosine or tryptophan or phenylalanine residues. The spectrophotometry is used to determine the concentration of proteins or other molecules that absorb light in the UV spectrum.

ii. Circular dichroism (CD)

iii. Fluorescence–tyrosine, tryptophan, phenylalanine are fluorescent

iv. NMR – uses nuclei of isotopes with a non zero spin (^{13}C ^{1}H ^{31}P ^{15}N) are magnetic and absorb electromagnetic energy

v. MRI

Synthesis Of Amino Acid Derivatives

I. Amino acids are important compounds in the overall survival of humans. Many of the amino acids are available through dietary intake. There are however 11 amino acids that are not available through the dietary intake and therefore must be synthesized metabolically. The overwhelming majority of the nitrogen in the world is found within the chemical structure of amino acids and in nucleotides.

A. Alanine must be synthesized from the precursor pyruvate. This is accomplished through a transamination reaction.

B. Aspartate is synthesized through a transamination reaction of Oxaloacetate.

C. Glutamate is synthesized from the precursor α keto-gluterate. This is through a reductive amination reaction.

D. Arginine is synthesized as a byproduct of the urea cycle.

E. Glutamine is synthesized by the amidation of glutamate.

F. Proline is synthesized from 3 phosphoglycerate of the glycolysis reactions.

G. Glycine is synthesized from serine.

H. Cysteine is synthesized from methionine and serine.

I. Tyrosine is synthesized from phenylalanine.

II. Glucogenic amino acids are those amino acids that can be used in gluconeogenesis. The glucogenic amino acids are Glutamate, Glutamine, Aspartate, and Alanine.

A. The reactions required to manufacture the glucogenic amino acids need energy input to run. The transamination reaction runs using glutamate as the main amino acid for the interconversions of the amino acids of this reaction pathway. The reaction is as follows:

Glutamate + ATP + NH_3 = glutamine + ADP + Pi with the enzyme glutamine synthase .

B. The enzyme that is responsible for this reaction is found in all cells, specifically in the mitochondrial of all cells. In the liver, the enzyme is found within a small population of paravenous cells. The activity of this enzyme is stimulated by the presence of oxogluterate. It should be noted that the conversion is between the 2 oxogluterate and the amino acid via this transamination. These transamination reactions are vital in both the synthesis and also the degradation of the amino acids and derivatives of these amino acids. The degradation of the glutamate is via glutamate dehydrogenase and is generally seen to occur within the liver.

(The reaction is : $=C\text{-}NH_2$ + $=CO$ ←→ $=CO$ + $=C\text{-}NH_2$). The catalyzed reaction is begun as a nucleophilic attack of the Schiff base structure found between the pyridoxal phosphate and the ε lysine group of the enzyme. The product of this reaction is the release of free ammonia and that ammonia is then found incorporated into the amino acid glutamine. If this reaction does not run, the free ammonia will quickly become toxic. (<60 μmol/l). Degradation of glutamate is the reverse of the synthesis yielding the 2 oxogluterate. The ammonia that is liberated will either be eliminated via direct excretion or by metabolic entry into the Urea cycle. The oxo – acids are able to be utilized in the gluconeogenic pathway once they are oxidized.

a. Glutamine is able to be incorporated into the developing polypeptide chain through the transammination reaction. The enzyme responsible for this reaction is transaminase. This reaction will bind the amino group of the glutamine to be the point of incorporation.
b. It is possible for the liberated ammonia to be used for energy once it has been acted upon by the enzyme glutaminase and converted to glutamate.
c. Glutamate is able to be interconverted into other compounds that are able to act as donors of the amino group in a variety of other reactions. Glutamate is able to used in the CNS as GABA 4-aminobutyrate, the decarboxylated product of glutamate. As the GABA is degraded, it will eventually form succinate.
d. Glutamate is also able to be converted to N-acetyl-glutamate through a transamination and condensation reaction with Acetyl CoA.
e. Glutamate is also essential in the ability of the blood to clot. Clotting requires a posttranslational

γ carboxylation. Vitamin K is required as a cofactor in this reaction.

C. Serine and glycine are products of reactions involving 3-phosphoglycerate. Serine and glycine are inter convertible amino acids. This interconversion requires tetrahydrofolate to participate in the reaction for the transfer at C_1.

a. Serine begins as 3 phosphoglycerate and is acted upon by a dehydrogenase that will yield 3 –P-hydroxypyruvate. Further reaction will be the transamination to 3 –P serine.

b. Serine is important as a component of glycerophospholipids. It is the originating amino acid for sphingosine and ceramide synthesis.

c. Serine is degraded into pyruvate through the action of the enzyme serine dehydratase. It is via this reaction and product that the two amino acids, serine and glycine are able to enter the gluconeogenic metabolic pathway.

d. The interconversion between these two amino acids is the main method of degradation of glycine. This reaction is dependant upon the C_1 transfer by tetrahydrofolate.

e. In the synthesis of glycine the methylene group from the serine will be transferred to tetrahydrofolate to produce 5,10 methylene – THF. The enzyme that is responsible for this transfer is glycine hydroxymethyltransferase and it requires pyridoxal as a cofactor.

D. The synthesis of Proline is originating from Glutamate. The reaction for this synthesis is the phosphorylation and reduction to glutamine semialdehyde. The glutamine semialdehyde is then nonenzymatically cyclized to L-1-pyrroline-5-carboxyate. The final reaction is the reduction to proline. This final reduction is NADH dependant or NADPH dependant reduction.

a. The physical properties of proline make it more rigid at the carboxylate –C-N bond. This property make proline inhibit the formation of α helicies.

b. This rigidity is also a reason for the difficulties of proline being incorporated into the process of formation of collagen.

E. The synthesis of lysine, methionine and threonine is seen to be initiated through modification of a secondary amino acid. The formation of these amino acids arises from the amino acid aspartate. In the formation of these amino acids the aspartate is phosphorylated.

F. The synthesis of lysine is as a condensation reaction and subsequent reduction of aspartete semialdehyde and pyruvate followed by the reduction of the product trimethyllysine.

G. Synthesis of Threonine is done via the isomerization of homoserine during the formation of a phosphate ester with the hydroxyl group. Once the isomerization is completed there is a final pyridoxal phosphate elimination via a β-γ elimination reaction to complete the synthesis. Threonine is converted by threonine dehydratase into 2-oxobutyrate that is a precursor of isoleucine. The degradation of threonine is that is it converted to glycine and acetylaldehyde or acetate and then further reacted on to produce acetyl CoA.

H. Cysteine synthesis is rather unusual in that it requires homocysteine to be converted into cystathionine by using the sulfur from homocysteine. During the degradation of cysteine, taurine is formed through the oxidation of the –SH and then subsequent decarboxylation.

I. Methionine is synthesized through a multiple reaction process that begins with the activation of the γ hydroxyl of homoserine. The activation of this gamma hydroxyl is through one of the high energy phosphate compounds, (succinyl, acetyl or other phosphate). The product of this reaction will yield cystathionine that will be converted to homocysteine. The homocysteine is then reacted upon in a methylation that will use methyltetrahydrofolate to finalize the reaction pathway and produce methionine. The formation of S adenylsyl-L-methionine or SAM is accomplished through a series of reactions that begins with S-adenylylation of the amino acid methionine. The product will be a positively charged sulfur with an active methyl group. The function of this reaction is important as a methyl donor in many biosynthetic reactions.

II. In normal metabolism, there are many transfers of single carbon groups during the metabolic processes. In order for these transfers to be completed, there must be a carrier for the carbon units. The carriers are of a variety of different types as the carbon units are of differing oxidation states.

III. The standard free energy that is associated with the phosphoryl transfer has been well described and documented. The reaction ATP + Pi + H^+ has a $\Delta G^{o'}$ of -7.3 kcal/mol where we see Glycerol 3-phosphate + H_2O ←→ glycerol + Pi has a $\Delta G^{o'}$ of – 2.2kcal/mol. These values define that the phosphoryl potential of ATP is greater than that for glycerol 3 phosphate.

AMINO ACID DERIVED COMPOUNDS

Fig 8

COMPOUND	PRECURSOR	PRIMARY FUNCTION
Dopamine	Tyrosine	Neurotransmitter
Epinepherine	Tyrosine	Hormone
GABA	Glutamate	Neurotransmitter
Histamine	Histidine	Vasodilator
Melanin	Tyrosine	Pigment
Melatonin	Tryptophan	Hormone
Norepinepherine	Tyrosine	Neurotransmitter
Serotonin	Tryptophan	Vasoconstrictor
Thyroxine	Tyrosine	Hormone

A. Biotin is the carrier for the Carbon Dioxide, CO_2. Biotin is considered to be the essential cofactor in the carboxylation reactions.

i. The beginning point of biotin synthesis is Malonyl CoA. Biotin is found to be covalently bound via an enzyme through an amide bond to the ε amino group of Lysine.

ii. Biotins main function is the transfer of carboxyl groups.

iii. The enzyme biotin carboxylase. This enzyme is responsible for the addition of the $HCO\text{-}_3$ to the biotin.

iv. The enzyme transcarboxylase is responsible for the addition of CO_2 via the biotin carrier protein. These enzymes are activated by the bicarbonate available and leading to the carboxylation of the biotin at the N1 position.

The reactions are as follows:

HCO^-_3 + ATP + biotin ←→ADP + Pi + biotin- CO_2 (Carboxylation)

Biotin – CO_2 + H_2O ← → Biotin + H_2 CO_3 (Decarboxylation)

Biotin - CO_2 + pyruvate ← → biotin + oxaloacetate (Transcarboxylation)

v. In humans the source for the biotin binding protein is generally found in egg whites and the yolks of eggs. The amount of biotin that is deemed as a requirement in adults is 200 μg/day.

B. Tetrahydrofolate is considered as the universal carrier as it can carry all but CO_2. The outstanding characteristic of the tetrahydrofolate is the inclusion of the pteridine ring. In humans this compound is required as a vitamin (B_c). Synthesis of the folates is from GTP being converted to dihydrobiopterine P_3 and then to a removal of all the phosphates and then pyrophosphorylation to 2 amino-4hydroxy-6-hydroxymethyl-7,8-dihydropteridine-P_2. This conversion will allow the addition of glutamate that will produce the dihydrofolate. This is a energy expensive reaction that will utilize ATP to drive the conversion. The final aspect of this reaction is the reduction of the dihydrofolate into tetrahydrofolate.

C. S – Adenosylmethionine is the major carrier of the methyl groups in the body. It functions as a donor of the methyl groups to many biochemical reactions.

D. Synthesis of catecholamines (dopamine, epinephrine, norepinepherine) is important as the derivatives are functional neurotransmitter and hormones.

i. Tyrosine will be hydroxylated to 3,4 dihydroxyphenylalanine. This reaction is catalyzed by the enzyme tyrosine hydroxylase. This is important as there is a cofactor required, tetrahydrobiopterin. The end result of this reaction is the formation of dopa, the precursor of dopamine. The reaction will continue as the dopa will be converted to dopamine through the aromatic acid decarboxylase to yield 3,4 dihydroxyphenylalanine (dopamine)

ii. The formation of norepinepherine is a continuation of the above reactions. The enzyme dopamine β hydroxylase will react with dopamine's side chain to hydroxylate it and yield norepinepherine. This reaction requires the inclusion of Cu, ascorbate, and molecular oxygen to be bound to the enzyme for activation. This is a very important reaction as norepinepherine has been demonstrated experimentally to be the major neurotransmitter of the sympathetic nervous system.

iii. Synthesis of melanins, the biologic pigments. This process is completed in the melanocytes. The synthesis is initiated with the conversion of the amino acid tyrosine into dopa. This reaction is different in the melanocyte than other cells as the enzymatic activity is from tyrosinase. What makes this different also is the next reaction in this pathway is catalyzed by the same enzyme in the conversion of dopa into dopaquinone which is then converted to melanin. This reaction requires a cofactor of copper.

iv. Synthesis of thyroxine and triiodothyronine begins with the conversion from tyrosine.

a. Thyroxine (T4) 3,3,5,5, tetraiodothyronine
b. 3,3,5 Triiodothyronine (T3) is the more active of the two
c. Both of these are synthesized in the thyroid gland follicle cell. This reaction is an iodization reaction of the tyrosine side chain. The iodinated form is stored within the follicle cell lumen until the time of secretion as T_3 will become converted to T_4 for release into the circulation.

II. Tryptophan metabolism for synthesis of serotonin, melatonin and NAD+.

A. Serotonin is the vasoconstrictor as well as promotes the contraction of the smooth muscle.

i. Additional important functional duties of serotonin is as a neurotransmitter in the hypothalamus and the brain stem. Serotonin will function as a neurotransmitter in the pineal gland and the chromaffin cells of the gastrointestinal tract.

ii. Degradation of serotonin is done via the enzyme monoamine oxidase that catalyzes conversion of 5 HT to 5 hydroxyindoleaceteladehyde.

B. Synthesis of melatonin that is synthesized in the pineal gland.

i. conversion is from the enzyme N – acetyl transferase that will convert 5 HT to 5 hydroxy –N – acetyl tryptamine.

ii. The 5 hydroxy –N – acetyl tryptamine is them methylated by the O methyl transferase and S adenosylmethionine to melatonin.

iii. It is important to note that the nicotinamide ring of NAD+ is synthesized from tryptophan.

C. Gamma aminobutyrate (GABA) is synthesized from the amino acid glutamate.

i. GABA functions as an inhibitor within the brain and spinal cord.

ii. It becomes metabolized by the neurons to succinate of the TCA.

D. Histidine is the precursor of histamine, a very potent vasodilator.

i. Histamine is also a neurotransmitter and a mediator of the immune reaction.

ii. The synthesis is catalyzed by the enzyme histidine carboxylase.

E. Glutathione is a tripeptide that is formed from the amino acids glutamate, cysteine, and glycine.

i. This molecule can exist in two forms a monomeric form and an oxidized form.

ii. particular to this molecule is the peptide bond that it includes, there is a peptide bond formed from the γ carboxyl and not the α carbon.

iii. the function of glutathione is as a sulfhydryl buffer. It also serves as a transport mechanism across the cell membrane for other amino acids. Lastly, it functions as a cofactor for glutathione peroxidase that utilizes the reduced glutathione to detoxify peroxidases.

F. Synthesis of creatine phosphate compounds begins with parts of three amino acids, glycine, arginine and S adenosylmethionine.

i. Creatine is a very high energy phosphate donor.

ii. Synthesis is from the transference of the guanidium group of arginine to glycine with the end product being guanidinoacetate that will itself be acted upon by S adenosylmethionine to finally yield creatine.

iii. ATP is used to provide the phosphate

iv. Urinary creatine is a reliable measure of the body cell mass via the relation of the 24 hour excretion of creatinine to the individuals height.

v. Blood creatinine levels are also a reliable indicator of kidney function. This is due to the clearance pathway being serum to the kidney to urine formation.

III. Heme is a prosthetic group that is found on a variety of enzymes and other proteins. In Hemeproteins the heme molecule may serve several functional roles. In cytochromes, heme functions in the capacity of transferring electrons. In this situation, the heme is capable of being oxidized and reduced. In the enzyme catalase, the function of heme is very different. In this capacity it assists in the breakdown of hydrogen peroxides. Then in myoglobin and hemoglobin the heme molecule is given the function of binding to oxygen. It is composed of 4 pyrrole rings that constitute the porphyrin ring. The heme is maintained within the center of the ring via the binding of 4 nitrogens that are in the ring itself.

A. precursors of heme are glycine and succinyl coenzyme A (succinyl CoA).

B. In its role of being a prosthetic group, heme is associated with Hemoglobin, Cytochrome C, catalase, and peroxidases.

C. The synthetic pathway is as follows:

   i. δ –Aminolevulinate that is derived from the amino acid glycine and succinyl CoA that is catalyzed by the enzyme δ aminolevulinate synthase. This enzymes activity is inhibited by increased levels of lead. The reason for this is that pyridoxal phosphate is the cofactor.

   ii. Two molecules of δ aminolevulinate will join to form prophobilirubin by the action of the enzyme δ aminolevulinate dehydratase.

   iii. Next step requires 2 enzymes for the reaction to run, uroporphyrinogen synthase and uroporphyrinogen III cosynthase. In this step there will be formation of a tetrapyrrole made from 4 molecules of prophobilinogen.

   iv. Synthesis of protporphyrin IX requires 3 enzymes that will act in the conversion from uroporophobilinogen.

a. Uroporphyrinogen decarboxylase is responsible for the decarboxylation of 4 of the side chains of uroporphobilinogen II to yield coproporphyrogen III.
b. The next two reactions are linked in end product as the first will be Coproporphyrinogen oxidase as it will decarboxylate 2 of the side chains coprorphyrinogen III as it produces protoporphyrinogen IX.
c. Protoporphyrinogen oxidase will remove 6 hydrogen atoms in the production of the final product of protoporphyrin IX.

v. Protoheme IX (heme) is the product of the insertion of the Fe through the activity of ferrochelatase.
   i. As mentioned above, the nitrogen molecules in the ring bind to heme and maintain the heme in its position.
   ii. In the case of the oxygen binding functional capacity of myoglobin and hemoglobin, the additional bonds of the ring that are created from the Fe^{2+} are replaced with histidine.
   iii. Methemoglobin is produced from the oxidation of the heme of the protein myoglobin to the ferric form of metmyoglobin. The same is true for hemoglobin where the heme is oxidized to the ferric state as well producing methemoglobin. This ferric state prevents the binding of oxygen and allows water to bind in its place.
   iv. Myoglobins role in physiology is to bind and carry oxygen for delivery to the muscle. It is

found within the muscle tissue. Structurally, it is a single peptide very similar to the peptide chains of hemoglobin.

a. This is a very compacted molecule. 80% of the polypeptide chain is folded into 8 alpha helical regions, A-H. Structurally, these regions are known to terminate with either a proline or β bend. This configuration is stabilized by the hydrogen and ionic bonds of the polypeptides.

b. The heme of myoglobin is situated in the hydrophobic pocket with 2 histidine residues within the structure. These are called the proximal and the distal histidine. These histidines have very different roles within the overall function of the molecule. The proximal histidine is responsible for the direct binding of the heme molecule through the histidine side chain. The distal histidine is very unusual in that is does not directly impact the heme molecule rather, it creates stability on the oxygen as it binds to the heme molecule. Myoglobin is only able to bind a single oxygen since it has only a single heme molecule.

v. Hemoglobin as opposed to myoglobin is found only within the erythrocyte. It has only one function in the erythrocyte, that of carrying oxygen to the cell via the capillaries and carbon dioxide from the cell to the lungs for exchange.

a. The structure of the hemoglobin molecule is a tetramer. There are 2 identical dimers (αβ)1 and (αβ)2 that are held tightly by the hydrophobic interaction between the nonpolar residues within the dimer. It is this tetramer that allows for hemoglobin to bind 4 oxygen molecules. It is also an additive mechanism for the binding of the oxygen molecules. The first is the most difficult to bind and then the remaining three each bind with greater ease.

b. There are 2 major states of the hemoglobin molecule designated "T" and "R". These designate the tense and the relaxed forms of the dimers. In the T form, the molecule is called deoxy or the low affinity form. In this configuration, the ability of the dimers to let oxygen releases is very strong. In the R form, the affinity of the molecule for oxygen is high. There is definitive evidence of some ionic bonds that will be broken for the oxygen to remain tightly held.
c. The binding of carbon dioxide will stabilize the T form of hemoglobin.
d. The bulk of the carbon dioxide which is produced in the body becomes hydrated and as such will be transported as bicarbonate ions. There will be a small percentage that will not hydrate and will form carbamate and transported with the uncharged alpha amino.
e. Once this compound is transported to the lung, it will be exchanged in the alveolar capillary for the oxygen molecules.

IV. Urea cycle is the main form of nitrogen disposal of human metabolism. In this cycle, the end product of urea has one of the nitrogens donated by free NH_3 and the other nitrogen is donated from aspartate. The CO_2 that is the substrate of the Urea cycle will donate the carbon and the oxygen of Urea. The urea cycle has two organs that work in unison to handle the Urea cycle. The actual formation of Urea is done in the liver while the excretion of urea is via the urine so the liver will produce the Urea and pass it to the kidneys for excretion.

A. This reaction pathway is in many instances much like the oxidative phosphorylation in that the pathway in found in the mitochondria. The initial two reactions of the Urea Cycle are found within the mitochondria. The remaining reactions of the pathway are found in the cytosol of the mitochondrial.

The accounting of this reaction is:

Aspartate + NH_3 + CO_2 + 3 ATP→ Urea + Fumarate + 2ADP + AMP + 2 Pi + PPi + 3 H_2O

B. Carbamoyl phosphate synthase I is the enzyme that is found to produce the carbamoyl phosphate. This is a very energy expensive reaction. The co factor required for this is N-acetylglutamate. There is also a second enzyme required for this path, carbamoyl phosphate synthase II that is found in the cytosol. This step of the reaction pathway does not require the co enzyme N-acetylglutamate.

C. The remaining reactions are oxidative deamination that will liberate the amino group as a free ammonia. Further metabolism of these products will yield α ketoacids and additional ammonia. The majority of the oxidative deamination will be of glutamate. The oxidative deamination of glutamate will require the enzyme glutamate dehydrogenase. This enzyme is unusual in a number of ways. First it can utilize either NAD+ or NADP+ as a co-enzyme. Second, it can react on glutamate very rapidly.

D. Citrulline is formed from the loss of the high energy Pi from the carbamoyl phosphate that will produce Citrulline. Citrulline will be the compound that is able to be transported across the mitochondrial membrane and into the cytosol.

E. Arginosuccinate is the product of the condensation reaction with aspartate. This is an important reaction since this is the step that yields the second nitrogen of urea. In this reaction, the energy expenditure is great as well with the ATP cleaved to AMP and PPi. Arginosuccinate will be reacted upon to yield arginine and Fumarate. Fumarate will become hydrated to form malate. Once this step is completed, malate can be transported across the mitochondrial membranes and reused in the TCA cycle. The cytosolic malate will be converted to glucose or to aspartate. The end product is dependant on the intermediate step passing through Oxaloacetate.

F. Arginine will metabolize to ornithine and urea. This is a very special reaction since it can only take place in the liver. The liver is the only place where this cleavage can take place. Synthesis is able to be done in other tissues, but the cleavage is localized to the liver.

G. Urea will diffuse to the kidneys from the liver. Urease is an enzyme that can cleave the urea from the liver and produce CO_2 and NH_3. Some of the ammonia that is produced in this step will be sent to the feces for excretion and the remaining ammonia sent to the blood again. This resorption can become pathologic quickly.

Fig 9

Protein
Amino Acids
Alpha Ketogluterate
aminotransferase
alpha ketoacids
Glutamate
NAD+
oxaloacetate
glutamate dehydrogenase
aspartate aminotransferase
alpha ketogluterate
alpha ketogluterate
NH3
Aspartate
CO2
carbamoyl phosphate
citruilline
ornthinine
arginosuccinate
arginine
fumarate
Urea

Enzymes

I. Properties of Enzymes

A. Enzymes are not used or produced during their use in a reaction that they catalyze.

B. Enzymes only facilitate a reaction; they do not elicit a reaction to run. They do not alter the equilibrium of the reaction.

- A century ago only a few enzymes were known, and virtually none were understood or characterized
- Now many thousands known, many well characterized. Many have similar functions, needed a system to distinguish between them
- International Union of Biochemists (IUB) developed a specific nomenclature and numbering system to classify them, avoid confusion
- Placed all enzymes in 6 separate classes based on what chemical groupings they catalyzed and how they did it.
- To be utilized in the body, a molecule must be absorbed then distributed through the body and then be able to penetrate the cell. This ability to penetrate is variable with regard to the mechanism, possibly simple diffusion, active transport, co-transport facilitated diffusion. This requires that the molecule be able to be distributed in its active form or able to be transported in inactive

form and then converted once in the cell to active form. These requirements are demonstrated to be best achieved by small molecules, generally less than 500 Daltons and are generally hydrophobic with heteroatoms and hydrogen donors.

- Conversely, the best targets for a drug would then posses small molecular weight, larger number of hydrophobic atoms that are available for binding through the creation of a polarized pocket.

C. Enzymes are most often proteins. Often RNA can act like an enzyme and if this is the scenario, the RNA is called a ribozyme. It will exert its effect through the manner it is able to catalyze the synthesis or the catalysis of certain phosphodiester bonds.

D. Enzymes are very highly specific function. They produce very expected products. These expected products as well as the enzyme itself are the basis for molecular targeting in pharmacology.

- There is abundant data that suggests the human genome is filled with enzymes that are able to attack the encoding sequences of the so called "drug gable targets". It is possible that a protein will contain a sequence that can be considered as a "drug gable target" but yet not be ideal as a drug. This is evidenced in the nearly 30,000 proteins that are known to be encoded within the human genome, and only approximately 3,000 of those same sequences can be effective as a drug or have a positive effect in regards to the disease.

- In pharmacology, oral administration of a substance is the most desirable method of delivery and enzymes are ideal in this role.
- Enzyme inhibitors are also therefore ideal targets as well.

E. It is possible that an enzyme may work with more than one substrate. However, the specificity is determined by the functional groups (substrate, product, enzyme itself, and cofactors), the actual physical proximity. Actual conformation of the enzyme – substrate complex as it fits naturally, (lock & key) or induced conformational changes by molecular alterations.

- The ability of the enzyme to exert its effect is due to the capacity of that enzyme to interact with the target molecule. This occurs through the active site within the enzyme.
- The active site of the enzyme is very small in comparison to the overall size of the enzyme. In order to prevent the bulk attachments of substrate to the enzyme, the active sites are usually located within a cleft. This positioning of the active site will enhance the specificity of the enzyme. The activity of some of the enzymes will influence the activity as well. Peptidases that hydrolyze peptide bonds are excellent examples of this enzymatic action.
- Additional factors that will influence the effectiveness of the interaction will be any cofactors, metal ions, other metabolites, or the actual physical structure of the polypeptide as

steric hinderance may create a problem for the interaction desired. These properties also make them targets for inhibition via chelation (captopril & enalapril) and thus altering the binding capacity of the enzyme.

- In most instances, the initial interaction between the enzyme and the substrate will be a non covalent interaction. This is the reason for the availability of the hydrogen donors to be available so that there can be the interaction and possible donation of those hydrogens. Other aspects of chemical bonding that are important are the hydrophobic interactions, Van der Waals attractions and electrostatic attraction or repulsions. Enzymes are able to achieve multiple configurations as are the substrates and thus the importance of the chemical activity of the pocket of the active site becomes more imperative. This property of enzymatic activity is demonstrated in the nucleoside analogue inhibitors and the non nucleoside inhibitors of HIV reverse transcriptases.
- The pharmokinetic activity of drugs / drug metabolism is influenced by the properties of these changes in conformational states, or biotransformation. Elimination of many xenobiotics from the body are influenced by the transformations that occur secondary to these biotransformation of enzymes.
- These reactions are described as Phase I or Phase II reactions. In phase I reactions, there

is increased aqueous solubility of a compound that makes it easier to eliminate that drug. This is accomplished via the metabolism of the parent compound into a more polar byproduct through an oxidation reaction, a hydrolytic reaction or a reduction reaction.

F. Enzymes work in a specific range of temperature and pH.

G. Free energy changes are seen as the value of the free energy of the substrates.

H. Energy of Activation is decreased through the use of enzymes. In order to have this happen, there is a transition state of the enzyme substrate complex.

I. Enzymes are protein catalysts that will increase the actual rate of the reaction while themselves not being used in the reaction.

II. Enzyme function is through the method of pathway alteration for the change to a more energetically favorable pathway.

A. The free energy of activation is the expression of the difference between the reactants and the high energy intermediate that occurs in the product formation.

B. Mathematically, the free energy of activation is a transitional state of the kinetics of the reaction during the conversion of reactants to product. Enzymes are destined to reduce the energy of activation.

Fig 10

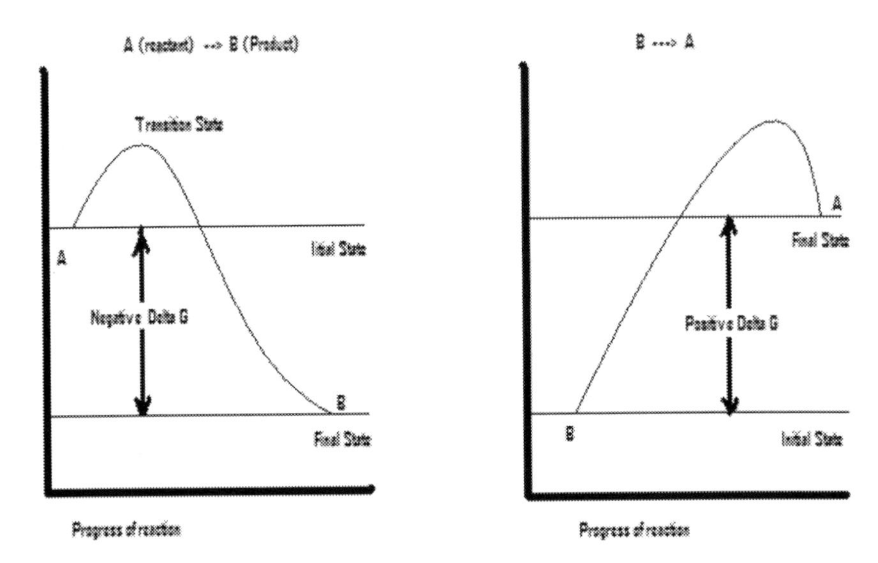

C. In the reaction of A↔ T↔B, the molecules of "A" must contain enough energy to overcome the transitional state and proceed to become "B".

1. In this reaction the lack of enzymes means that every molecule must inherently contain the energy to overcome the transitional state. This means that there likely will be few molecules that will contain enough energy to overcome this transitional state. Therefore, the rate of the reaction will be very slow.
2. The lower the energy of activation, the more molecules able to overcome this energy state.
3. Enzymes will lower the energy of activation and therefore, increase the reaction rate.
4. The enzyme does not change the free energies of the reactants or of the products. In this manner they do not alter the equilibrium of the reaction.

D. The enzyme will combine reversibly with the substrate to form the enzyme substrate complex [ES]. All enzyme based reactions will be initiated by formation of a encounter complex that exhibit's a binary behavior between the enzyme and the substrate.

1. Michaelis – Menten equation is the expression of the velocity to substrate ratio.
2. Vo = Vmax[S] / Km + [S] , for this equation to be valid, there are some assumptions that need to be made. The concentration of substrate[S] must be greater than the concentration of enzyme [E] . The rate of synthesis and breakdown of the [ES] complex is equal.
   a. Km is a reflection of the affinity of the enzyme for the substrate. This value represents the substrate concentration where the reaction velocity is ½ Vmax. Km is not variable with the [E].
   b. Low Km values indicate a strong affinity for the enzyme towards the substrate. Therefore, a small concentration of the substrate is required to reach ½ Vmax.
   c. High Km values indicate that there is a weak affinity for the enzyme towards the substrate and therefore, the amount of [E] required will be large.
   d. A first order reaction is when the [S] < Km, making the velocity of the reaction proportional to [S].
   e. A reaction is termed a zero order reaction when the [S] > Km, the velocity of the reaction is now at a constant and it is now equal to the Vmax.

In this reaction type the rate of the reaction is independent of the [S].

f. Although most enzyme reactions are reversible, some directionality is imposed on particular steps, especially where PPi is one of the products:
   A+B+ ATP ◊ A-B + AMP + PPi
g. The PPi rapidly decomposes to 2 Pi, which effectively prevents the reverse reaction because one of the potential substrates (PPi) is no longer present
h. Enzyme catalyzed reversible reactions may be catalyzed by the same or a different reaction
i. Forward reaction gives rise to K1, reverse reaction to K2 or K-1
   K1 S <-> P K-1 or K2
j. Reversibility often based on [S] substrate concentration and tend to go downhill from higher to lower concentration.

III. Enzyme types

A. Oxidoreductases – oxidation reduction reactions
   a. Oxidases – use oxygen in the reaction but do not incorporate into the product.
   b. Dehydrogenases – use molecules other than oxygen as electron acceptor
   c. Oxygenases – place oxygen in the substrate
   d. Perioxidases – use H_2O_2 as electron acceptor

B. Transferases – transfer reactions, they transfer functional groups.
   a. Methyltransferases

b. Aminotransferases

c. Kinases

d. Phosphorylases

C. Hydrolases – transfer water

a. Phosphatases – remove PO_3- from substrates.

b. Phosphodiesterases

c. Proteases – cleave amide bonds

D. Lyases – will add or remove water, ammonia, carbon dioxide, or double bonds.

a. Decarboxylases

b. Aldoases

c. Synthases

E. Isomerases – create changes within the same molecule

a. Racemases

b. Mutases

F. Ligases (synthetases) – join molecules together and require the use of ATP doing so.

a. Carboxylases

b. Synthetases

IV. Kinetics of Enzyme reactions

A. The formation rates of product – enzyme complexes are proportional to time.

B. The formation rates of product – enzyme complexes are directionally proportional to the enzyme concentration.

C. Zero Order reactions are where the velocity or rate is constant and it is independent of the reactant concentration.

D. First Order reaction is directly proportional to the reactant concentrations

E. Second Order reactions are proportional to the product of the concentrations of the reactants.

F. The effect of the enzyme concentration on the reaction velocity is dependent on the stability of the concentration of the substrate. If this is held constant the velocity is proportional to the concentration.

G. The effect of the substrate concentration on the reaction velocity is first order if the enzyme concentration is low. If the enzyme concentration is mid level, the reaction is variable between zero and first order. If the enzyme concentration is high then the reaction will follow zero order kinetics.

H. Km for a reaction is the substrate concentration where the velocity is half of the maximal rate available. It is a measure of affinity, the lower the Km the greater the affinity. There is a relationship between the values of ΔG° and K_{eq}. In the reaction $A \rightarrow B$ there can be a point of equilibrium of the reaction where there will be no net chemical change of $A \rightarrow B$ or of $B \rightarrow A$. This point is defined mathematically by the equation, $\Delta G^\circ = - RT \ln K_{eq.}$

V. Enzyme inhibition is achieved through competitive inhibition, uncompetitive inhibition, and non-competitive inhibition. (Reversible) and Affinity labeling, mechanism-based or suicide inhibitors, transition state analogs (Irreversible competitive inhibitors)

1. Competitive inhibition has competition for the binding site. The inhibitor will bind reversibly to the same site as the enzyme binds. The effect on the Vmax is that the effect of the inhibition will be reversed if you increase the [S] substrate. The effect on the Km with competitive inhibition will be that the Km will increase significantly as there will be more substrate required to come to ½ Vmax.
2. Noncompetitive inhibition binds only to the enzyme substrate complex at a point other than the binding site. So the additional binding is done at the allosteric site.
3. Noncompetitive inhibition has the binding occur to both the free enzymes and the Enzyme substrate complex at the allosteric site. The effect of noncompetitive inhibition on Vmax cannot be overcome by increasing the [S], therefore it lowers the Vmax of the reaction. There is no effect on the Km due to the lack of interference of the inhibitor on the binding of the substrate to the enzyme.
4. Affinity Labels are substrate analogs that have a highly reactive group not seen on the natural substance. This blocks the active site permanently
5. Mechanism based or suicide inhibitors are substrate analogs that become activated by the catalytic action of the enzyme. Inactivated enzymes are the result of this type of inhibition.

6. Transition state analogs are very close to the natural transition state but they bind the enzyme tightly and do not release it.

VI. Enzyme regulation can be accomplished via changes in pH, temperature, covalent modification.

A. pH changes can denature protein, and since most enzymes are proteins, denaturing agents will alter enzyme function

B. Temperature will also denature proteins and as above will alter the function of that enzyme and its function.

C. Product accumulation will produce a negative feed back loop and inhibit the production of more enzymes.

D. Covalent modification is via phosphorylation, hydroxylation. Many enzymes are able to be regulated through the phosphorylation or dephosphorylation of a specific site on the enzyme. This in effect will "active" or "deactivate" the enzyme.

i. In phosphorylation and reactions, ATP is the donor for the phosphorylation reactions. For this phosphorylation to be done, protein kinases are required as the enzymatic catalyst.

ii. In reaction s where there is a cleavage of a phosphate, a dephosphorylation reaction, the removal of the phosphate if through the enzyme phosphoprotein phosphatases.

E. Enzymes are also sensitive to the [S] since many of the substrates that are found within the cell are at or close to the values established for Km. Enzymes are also able to be regulated by the cell itself via the cells ability to synthesize the enzyme. This ability is called induction or repression of the enzyme.

F. Enzymes are regulated as mentioned above through allosteric regulators. These allosteric regulators are usually molecules that are celled effectors. Effectors are able to bind to the receptor site. It is important to understand the bond type that is formed in this receptor –substrate complex. This bond has enough energy to create conformational changes in many enzymes, but not posses such large energy amounts that it is not able to be released at great energy expense. The greatest point of interest in this type of regulation is that the receptor site that is being discussed, is not at the active site for the enzyme. It is possible to be located on a different domain of the protein. If the allosteric regulator indices an inhibitory effect, then it is termed a negative effector. The opposite is found to be true then, if the allosteric regulator induces an increase in the activity of the protein function, it then is a positive effector.

i. If the allosteric regulator is the effector then the enzyme is said to induce a homotropic effect.
ii. The most common function of this reaction regulation is as a positive effector.
iii. It is possible that the binding of the allosteric regulator will induce a cooperative binding effect. This is analogous to the binding of oxygen as it binds to the heme. The initial oxygen molecule

is the most difficult to bind then each successive molecule of oxygen binds easier. In this design, the binding of the allosteric activator will induce a more efficient catalytic activity at distant enzyme – substrate complexes. Mathematically, this is expressed as the sigmoidal curve that will be plotted as you compare the reaction velocity (Vo) :[S].

iv. In the situation where the effector is not the substrate, the reaction produces a heterotrophic effect. In this reaction design, the binding of the ES complex will induce another enzyme to activate and create a substrate or effect that will have regulatory effects on the initial ES complex.

Hormones

I. Hormones will direct and coordinate metabolism throughout the body. They regulate through transmission of sensor information to target cell. The target cell is the cell that is destined to respond to the changes sensed by the receptor cells and directed through the hormone.

   A. Endocrine hormones are synthesized within the endocrine gland and then transported in the blood to a distant target cell.

   B. Paracrine hormones are synthesized locally with respect to the target cell.

   C. Autocrine hormones are unusual in that they exert the effect on the actual cell that synthesized the hormone.

   D. Hormones can be made of a variety of chemical substances

      i. proteins or peptides such as insulin that are actually secreted as larger macromolecules that require modification.

      ii. amino acid derivatives as in the thyroid hormones or catecholamines. These molecules are made from usual amino acids.

      iii. fatty acid derivatives such as the eicosanoids.

      iv. cholesterol derivatives like the steroid hormones, or sex hormones.

E. Hormones are classified by the solubility of the compound in water.

i. Hydrophilic hormones require binding to a receptor on the exterior of the cell in order to elicit a response within the cell as the binding occurs.

ii. Lipophilic hormones require the binding be to an intracellular receptor that will elicit a delayed response. The delay is due to the mechanism of action being transcription of genetic expression and not direct implications.

II. Lipophilic hormones are generally transported in the bloodstream via plasma proteins. The mechanism of uptake within the cell itself is diffusion. Within the cell there is a receptor that the hormone will bind to and create a structural change in the receptor that is the signal for the cell to produce the response.

A. The disadvantage of this hormone type is that there is a delayed time of response.

B. The advantage of this hormone type is the length of activation. The time of activation or the duration of the activation is hours to days. This is a tremendous benefit for a variety of cellular responses.

C. The receptors for the lipophilic hormones are generally proteins that will create the receptor hormone complex. These proteins are generally of a complex nature, having differing cellular domains. One of the domains will be responsible for the interpretation and recognition of the DNA of the hormone. This DNA sequence verification unit is called the hormone response element. The other domain will act as the receptor for the specific protein. This site is very specific for each hormone that it allows to bind. The typical description is the "lock and key" mechanism.

III. Hydrophilic hormones interact with the surface of the cell. Once the cell receptor recognizes the hormone it allows the binding to occur. The responses that are elicited are as specific as the hormone itself.

A. The structure of the receptor molecule is a large macromolecule that is anchored within the cell membrane. This protein has a very specific recognition sequence and therefore the receptor is very specific.

B. The binding of the hormone to the receptor is a reversible process.

C. The binding efficiency will decline as the serum concentration of the serum hormone declines.

D. This subset of hormones is able to elicit its action and response without actually entering the cell. This typically translates into a faster response that the lipophilic hormone but also a shorter time of activation.

E. The effects of many hormones are regulated by a special protein. This protein is the G protein. The unique aspect of this modulator is that it binds on the cytosolic side of the membrane.

   i. The "G" is designated because of the guanine nucleotide that is the backbone of the protein. It is able to utilize either GTP or GDP for protein binding.

   ii. The "G" proteins will have 3 subunits designated, α, β and γ. Specific function of the subunits will be as follows: α subunits will bind GTP or GDP. The β and the γ subunit will not bind to the nucleotide but rather to the α subunit directly. The active G protein is the Gα – GTP subunit.

iii. The G protein in its active state has further subdivisions as well.

a. G_α is the designate for the stimulation of adenylate cyclase

b. G_i is the designate for the inhibition of adenylate cyclase

c. G_{PLC} is the designate for the stimulation of phospholipase C.

F. Hormone receptors are possible to be directly linked to the G_α protein.

i. Adenylate cyclase is a large integral protein that functions in the formation of cyclic adenosine monophosphate from ATP.

a. The cAMP is a second messenger hormone. It exerts its effect from the binding of the hormone receptor complex to molecules within the cell.

b. cAMP will activate the enzyme protein kinase that will phosphorylate other cellular compounds.

c. cAMP is inactivated by the action of the enzyme cAMP phosphodiesterase.

d. glucagons and epinephrine will activate adenylate cyclase.

G. Hormones receptors can bind directly to G_i proteins that will inhibit the activity of the adenylate cyclase.

i. epinephrine will work in this capacity.

H. Hormone receptors can bind directly to the G_{PLC} protein to activate phospholipase C, a membrane bound enzyme. Phospholipase C will act to hydrolyze phosphatidylinositol 4,5 bisphosphate (PIP_2) to yield inositol 1,4,5, triphosphate (IP_3) and Diacylglycerol (DAG), both of which are second messengers.

Glycolysis

I. Glycolysis is a biochemical process that involves the degradation of carbohydrates into molecules of glucose. This is an important pathway as almost any sugar can be converted to glucose. This biochemical pathway continues the degrading of the glucose molecules into molecules of pyruvate. This particular pathway is composed of a series of nine reactions.

Fig 11

ATP↓

Glucose →hexokinase Glucose 6 Phosphate

↕***phosphoglucose isomerase***

Fructose 6 Phosphate

ATP→↕***phosphofructokinase***

Fructose 1,6 bisphosphate

↓aldolase

Glyceraldehyde 3 Phosphate

triose phosphate isomerase

↕⟶⟶ Dihydroxyacetone phosphate

Glyceraldehyde 3 phosphate
dehydrogenase

1,3 bisphosphoglycerate

ATP←↕ ***phosphoglycerate kinase***

3 phosphoglycerate

↕***phosphoglycerate mutase***

2 phosphoglycerate

↕***Enolase***

Phosphoenolpyruvate

ATP←↕***pyruvate kinase***

Pyruvate ↔ Lactate

A. Glycolysis is possible to be aerobic or anaerobic. The final outcome is still the same, the production of ATP.

B. Anaerobic glycolysis will yield 2 moles of ATP per mole of glucose.

C. Aerobic glycolysis will yield 6 moles of ATP per mole of glucose.

II. Staging of Glycolysis

A. The first stage of glycolysis involves the splitting of a 6 carbon sugar into 2 three carbon compounds. This process requires hydrolysis of 2 moles of Adenosine triphosphate per mole of hexose that is split.

B. The second stage of glycolysis involves the digestion of the three carbon molecules into molecules of pyruvate. This stage of the biochemical pathway will yield adenosine triphosphate rather than expend it. In this stage there will be 2 molecules of ATP generated for each molecule of pyruvate. The series of reactions will generate 2 molecules of pyruvate thus the total amount of ATP generated will be 4 molecules of ATP per molecule of hexose that enters into the pathway.

III. Glucose is the primary form of carbohydrate in the cells. It is derived from lactose (glucose & glactose) or from fructose (ketose isomer of glucose).

A. Glycogen is the primary form of carbohydrate storage in the human body.

B. Carbohydrates are metabolized into monosaccharides through the digestive enzymes within the digestive system.

i. Enzymes of carbohydrate digestion include: α amylase an enzyme that digests starch and glycogen in the saliva. There is an isozyme in the pancreas that is more important as it is responsible for the liberati on of this energy source. This enzyme hydrolyzes the α 1, 4 linkages of glucose. The limit if this enzyme is at the branching points of the glucose chain. This enzyme has no action on the α 1,6 linkages.
   The action of this enzyme will produce Maltose (α glucose) (1,4) glucose, maltotriose [α glucose (1,4) α glucose (1,4) glucose, and limit dextrins [highly branched molecule of 8 glucose joined by multiple α 1,6 bonds that cannot be degraded further]
ii. Oligosaccharidases that will complete the degradation of the disaccharides and the oligosaccharides on the epilithelial cells within the small intestines. This enzyme will remove from the non reducing end of the molecule [end opposite the ketone or the aldehyde].
iii. α Glucosidases within the small intestines is found to be plentiful. This is in stark contrast to the β glactosidases that digest lactose.

C. Transport of the products of the glycolytic pathway (D-glucose, D- glactose, D- fructose) requires active transport from the epithelial cells into the blood. The systems of transport are:

i. Na+ dependent monosaccharide cotransport system that is specific for D- glucose and D- glactose.

ii. Phlorhizin (plant glycoside) will inhibit the system and renal tubule transport of the monosaccharides as well.

iii. Na+ independent monosaccharide transport system will transport D- fructose via facilitated diffusion.

IV. Embden – Myerhof Pathway

A. Glucose to Glucose 6 Phosphate (G6P) is the phosphorylation of glucose and hexoses. This is done immediately upon the sugar molecule entering the cell.

i. Glucose is trapped within the cell since the phosphorylation of the sugar makes it very difficult to move through the cell membrane.

ii. The enzyme responsible is hexokinase with a molecule of ATP being hydrolyzed in this non -reversible step. The activity of this reaction is inhibited by the product, G6P. In the liver, this enzyme is glucokinase rather than hexokinase. Glucokinase levels are increased by carbohydrate rich meals and insulin. The main difference is that glucokinase is not inhibited by glucose 6 phosphate.

iii. This is an allosteric enzyme.

iv. Low Km for glucose, especially in the brain.

v. Glucokinase is found in the liver, and is the major phosphorylation enzyme of the liver. It is specific for glucose and has a high Km for glucose and therefore can handle a tremendous amount of glucose. It will increase activity in response to insulin and carbohydrate intake. This is not the situation for the hexokinase of the remaining tissues as it is not able to handle large concentrations of glucose.

vi. G6P is not only an important intermediate, but it is a precursor of several anabolic and catabolic pathways.

B. G6P is isomerized to fructose 6 phosphate. This is a reversible reaction and it is catalyzed by phosphoglucose isomerase, which isomerizes the C1 carbon. Glucose 6 phosphate is an aldose sugar that is isomerized to the keto sugar fructose 6 phosphate.

C. Fructose 6 phosphate (F6P) is converted to fructose 1,6 bisphosphate
(F 1,6 BP). This is a committed step and it is a non-reversible step. This is also the rate limiting step.

i. Enzyme responsible for this is phosphofructokinase (PFK 1) This is an irreversible reaction and the most important regulatory point of the pathway.

ii. PFK activation is done by the substrate itself, AMP that signals a state of low energy. Fructose 2,6 bisphosphate [formed from F6P when high levels are present will activate PFK in the liver only].

iii. PFK 2 is inhibited by citrate and ATP and is inhibited in its phosphorylated form. It is also inhibited by cyclic AMP (cAMP) – dependent protein kinase.

D. F 1,6 BP is converted to dihydroxyacetone phosphate (DHAP) and glyceraldehyde 3 phosphate (G3P) by the enzyme aldolase A.

   i. DHAP can be converted to G3P readily or converted to glycerol 3 phosphate. G3P is used as a substrate for the initial reaction of the second stage of glycolysis.

   ii. The conversion of G3P to DHAP is catalyzed by triose phosphate isomerase.

   iii. This is a very important reaction as the conversion of the DHAP to G3P will produce 2 units of G3P

E. G3P to 1,3 bisphosphoglycerate (1,3 BPG) is a reversible reaction that is catalyzed by glyceraldehyde 3 phosphate dehydrogenase.

   i. This is substrate level phosphorylation

   ii. Requires NAD+

   iii. This is a very energy intensive reaction, it will generate a high energy phosphate bond. This is because of the acyl phosphate (combination of phosphoric acid and carboxylic acid.)

   iv. This is the first oxidation reduction reaction of glycolysis.

F. 1,3 BPG converts to 3 phosphoglycerate (3-PG) is the first ATP generated in this reaction. This is substrate level phosphorylation and it is catalyzed by phosphoglycerate kinase.

   i. This is the initial production of high energy ATP. It is important to note that this reaction is going to produce 2 ATP per glucose molecule taken in during the initial step of glycolysis. This generation of the 2 ATP is due to the two molecules of 1,3 bisphosphoglycerate generated in the reaction.

G. 3 PG converts to 2 phosphoglycerate (2PG) in a reversible reaction that requires a cofactor (2,3 BPG) 2,3 bisphosphoglycerate. This is found in most cells but exceptionally high amounts in RBC's

i. This is a mutase reaction that is catalyzed by phosphoglycerate mutase.

ii. The 2,3 BPG is important for oxygen regulation and transport.

iii. 2,3 BPG is made from 1,3 BPG by a mutase with further degradation to 3 PG by a phosphatases.

H. 2 Phosphoglycerate converts to Phosphoenolpyruvate (PEP) is a reversible reaction that is catalyzed by enolase. The PEP is a high energy phosphate bond. This is a dehydration type of reaction that will yield the high energy phosphate compound.

I. PEP conversion to pyruvate is catalyzed by an enzyme complex called pyruvate kinase.

i. this enzyme is activated by F 1,6 BP.

ii. Activation is seen with increased concentration of high carbohydrate concentration and high insulin levels.

iii. Liver isozyme is regulated with covalent modification

iv. This enzyme is phosphorylated by cAMP-dependent protein kinase and it is in an inactive state when phosphorylated.

v. Dephosphorylation by a phosphatase will produce an activated form of the isozyme.

vi. Pyruvate kinase is inhibited by ATP, Alanine, Acetyl CoA.

vii. There are 2 moles of pyruvate that are produced from each molecule of glucose.

viii. This is also substrate level phosphorylation.

J. Pyruvate is converted to Lactate (lactic Acid) is converted by the enzyme lactate dehydrogenase.

i. This is the final product of the anaerobic glycolysis in the leukocyte, the cornea, the renal medulla and the erythrocyte.

ii. Lactate accumulation will drop the pH to a more acidic level with exercise. The increased ratio of NADH / NAD+ will drive the reaction to the reduction of pyruvate to lactate.

K. Pyruvate to Acetyl CoA is a non reversible reaction. This reaction is catalyzed by pyruvate dehydrogenase and it is accomplished in the mitochondria.

i. Acetyl CoA can enter the TCA and used as additional energy or used as substrate for the synthesis of fatty acids.

ii. There is an alternative metabolic pathway available in the metabolism of pyruvate and that is the carboxylation of pyruvate into Oxaloacetate. The enzyme that catalyzes this reaction is pyruvate carboxylase. This reaction is dependent on the cofactor of biotin. The final yield of this reaction is a substrate for the TCA.

L. Pyruvate to Alanine is a reversible and catalyzed by Alanine aminotransferase.

V. Adenosine triphosphate accounting

A. 2 moles of ATP per molecule of glucose are required to make fructose 1,6 bisphosphoglycerate.

B. 2 moles of ATP per mole of glucose come from the phosphoglycerate kinase enzyme

C. 2 moles of ATP per mole of glucose come from pyruvate kinase

D. there are 2 net moles of ATP per molecule of glucose produced via substrate level phosphorylation in anaerobic glycolysis.

E. In the anaerobic glycolysis the reaction accounting look like this

Glucose + 2 Pi + 2 ADP → 2 lactate + 2 ATP + 2 H_2O

F. In the aerobic glycolysis the reaction accounting looks like this

Glucose + 2 Pi + 2 NAD^+ + 2 ADP → 2 pyruvate + 2 ATP + 2 NADH + 2 H^+ + H_2O.

VI. Glucose is transported into the cells through very specific receptors and transport mechanisms. Glucose is unable to transport itself across the cell membrane via simple diffusion. In order for the glucose molecule to transfer into the cell there must be energy expended either in facilitated diffusion or in co- transport. There are 5 specific transport proteins for glucose, GLUT 1-5. These transport proteins are understood to demonstrate an active and inactive form. There is a phosphorylation of the receptor that will internalize the extracellular bound glucose to the internal of the cell.

A. GLUT 1 is plentiful in the red blood cells, AKA erythrocytes.

B. GLUT 4 is seen in larger concentrations in the adipose tissue and in skeletal muscle. These receptors are stimulated by insulin.

C. Co-transport is achieved through a carrier mediated coupling to the sodium channel. As sodium (Na+) is brought into the cell so is the glucose.

VII. Regulation of glycolysis is accomplished via the hormones glucagon and Insulin. Glucagon is a hormone that is produced in the Islet cells of the pancreas.

A. Glucagon will inhibit the conversion of glucose to glucose 6 phosphate while the reverse is true of insulin.

B. Glucagon will also limit the conversion of fructose 6 phosphate to fructose 1,6, bisphosphate. Insulin will have a stimulatory effect on this step of the reaction pathway.

C. Glucagon will limit the conversion of the conversion of phosphoenolpyruvate to pyruvate. Insulin is again a stimulator of this reaction.

Metabolism

I. All organism require energy for survival. There are many similarities as to the methods organisms use to derive the energy required for life. A sequence of organized biochemical reactions are called a metabolic pathway. There are 3 major categories of metabolic pathways: Anabolic pathways, in which new compounds are formed from simpler substrates, Catabolic pathways, in which substrates are biochemically degraded to simpler compounds, releasing reducing equivalents and Amphibolic pathways, which function as crossroads and link anabolic and catabolic pathways. Drugs use all 3 to impart their effects on metabolism.

A. There are two basic aspect of metabolism that are the basis of metabolism

i. Catabolism is the degrading of larger molecules into smaller ones. Catabolic reactions are going to degrade proteins, polysaccharides and lipids to end products such as CO_2, NH_3 and H_2O. The degradation of these macromolecules into the constituents is designed to capture energy. This energy is in the form of high energy compounds. Macronutrients will yield energy and the outcome of this catabolism will generate the following amounts of energy: CHO (4 kcal/g), protein(4 kcals/g), and fats(9 kcals/g).

Carbohydrates(CHOs): 4 kcal/gram are composed of simple sugars (sucrose, dextrose) and complex carbohydrates (starches) and fiber: (not digested in humans, therefore no kcals for human digestion. In truth there is about 1 kcal/g via colonic microorganisms. Proteins: 4 kcals/gram are termed as complete vs. incomplete depending on the side chain of the amino acids. Fats: 9 kcals/gram are either saturated(mostly from animals) or unsaturated(mostly from plants) fats. Alcohol: 7 kcals/gram is a poor source of energy but can at certain points become a very important source of energy.

ii. This process of catabolism has 3 stages.

a. Stage 1 is where macromolecules like TAG's, starches, and proteins are broken down into the monosaccharides, amino acids, glycerol etc. Specifically, the proteins are degraded into amino acids and polysaccharides are degraded into monosaccharides and the triacylglycerols degraded into free fatty acids and glycerol backbones. There is very little energy that is given in these initial steps.

b. Stage 2 is where those simple molecules are broken into smaller molecules that can be oxidized. The yield of these reactions will be acetyl Co A and a variety of other molecules. Some ATP is generated in this step.

c. Stage 3 is the final step in the role of metabolism. It consists of Citric Acid (TCA) (Kreb's Cycle), electron transport chain, Oxidative phosphorylation. These end up oxidizing things into acetyl coenzyme A (Acetyl CoA) and then into CO_2 and water. Lots of ATP is generated. The final common pathway of metabolism is the TCA cycle. In this reaction acetyl CoA will yield 2 moles of CO_2. There will be additional metabolic steps that will transfer 4 electrons to NAD+ and FAD that will yield NADH and $FADH_2$.

Fig 12

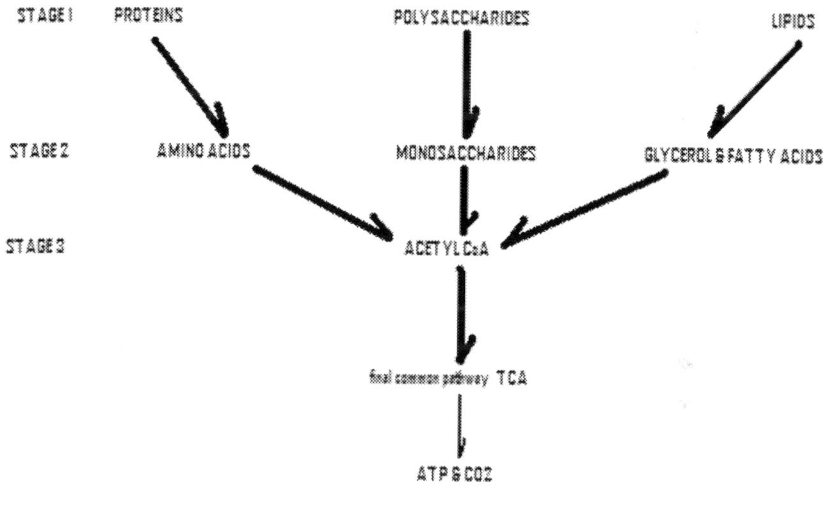

STAGE 1 PROTEINS POLYSACCHARIDES LIPIDS

STAGE 2 AMINO ACIDS MONOSACCHARIDES GLYCEROL & FATTY ACIDS

STAGE 3 ACETYL CoA

final common pathway TCA

ATP & CO2

Stages of Catabolism

iii. Anabolism is the process of building things from smaller things. Anabolic reactions are processes of divergence. In these reactions, there will be combinations of small molecules like amino acids that will form to make proteins that will form to yield fibrils that will yield fibers that will aggregate to form muscle. This type of reaction will require energy, often vast amounts of energy. This energy will typically be provided through the hydrolysis of ATP to ADP + Pi. There are additional requirements of reducing equivalents that are provided by NADPH.

iv. Amino acids make up the building blocks of proteins, and are stabilized by hydrogen bonds and linked together by peptide (C-N) bonds.

v. There are 20 amino acids that are essential for human metabolism, 8 of which the body can not make on its own

vi. The proportion of nutritionally essential amino acids generally decreases with increasing age, as it requires many more of the 'essential' to build new tissue than to maintain older tissues

B. The actual metabolic processes are found within the cells and within the specific organelles within the cell.

i. Cytosol contains glycolysis, Pentose phosphate pathway, and fatty acid synthesis

ii. Mitochondria contain the Citric acid cycle (TCA) (Kreb's Cycle) Oxidative Phosphorylation, Gluconeogenesis, Beta oxidation of fatty acids, and formation of Ketone bodies.

II. Energies

A. Standard State is defined as a pH of 7, 25 degrees Celsius, solutes at a concentration of 1 molar, and a pressure of 1 atmosphere.

B. Enthalpy (H) is heat work and defined as $H=E+PV$, Entropy (S) is the degree of randomness, Free energy (G) is the measure of useful work.

C. Free Energy (ΔG) is defined as the change on free energy during the synthesis of 1 mole of a compound under standard conditions.

D. Standard free energy is calculated by subtracting the $\Delta G^{0'}$ of the reactants from that of the products.

E. Reaction energies or the spontaneous nature of the reaction. This value is defined by the value of ΔG.

i. Exergonic reactions are able to proceed if there is a release of energy. In this type of reaction where the energy is released, it is secondary to the hydrolysis of a phosphoanhydride bond by addition a water molecule. In this reaction the product is Adenosine diphosphate and a liberated phosphate. This ADP +Pi is more stable than the ATP it is yielded from.

ii. Endergonic reactions are able to proceed if there is an input of energy.

F. Reactions can be coupled. This is when you have one reaction that is unfavorable being driven by a reaction that is favorable.

III. High Energy Compounds

A. High energy phosphate compounds have a high energy group potential. Defined this is a compound that has a highly negative $\Delta G^{o'}$ for the hydrolysis of the terminal phosphate group.

B. Adenosine triphosphate is the main high energy phosphate compound in most biologic organisms.

C. High Energy phosphate storage compounds are called phosphagens. Creatine phosphate is one of these compounds available in humans.

D. Other high energy phosphate compounds are:

i. Phosphoenolpyruvate (PEP) $\Delta G^{o'}$ -14.8kcal/mol

ii. Carbamoyl phosphate $\Delta G^{o'}$-12.3 kcal/mol

iii. Creatine phosphate $\Delta G^{o'}$ -10.3 kcal/mol

iv. Adenosine triphosphate $\Delta G^{o'}$-7.3 kcal/mol

v. Glucose 1 phosphate $\Delta G^{o'}$-5.0 kcal/mol

vi. Glucose 6 phosphate $\Delta G^{o'}$ -2.2 kcal/mol

E. Guanosine triphosphate is used in protein synthesis.

F. Cytidine triphosphate is involved in lipid synthesis

G. Uridine triphosphate is involved in polysaccharide synthesis

IV. Oxidation and reduction reactions involve the loss and the gain of electrons

A. Oxidative reactions involve the loss of electrons

B. Reduction reactions involve the gain of electrons

V. Electron Carriers

A. In aerobic organisms, molecular oxygen is the prototype electron acceptor

B. NAD+ and FAD+ are the carriers of majority of the electrons being pushed along the system. In NAD+ the nicotinamide ring is the acceptor, taking a proton and 2 electrons in the form of a hydride ion. While alloxazine ring is the acceptor in FAD+ accepting 2 protons and 2 electrons, or 2 hydride ions.

C. NADPH is the donor of the electrons and the hydrogen in reductive reactions.

VI. Regulation of Metabolism can be done from signaling within the cell or from communication between the cells.

A. Signaling from within the cell is via the substrate availability, substrate concentration, product inhibition, or allosteric inhibitors and or activators.

i. In this system, there are 2 main mechanisms of action for this signaling. In one system, the signal transduction is translocated into the cell by a receptor in the nucleus or the cytoplasm. This receptor –agonist complex will move through the nuclear pore to be bound to the enhancer region of the DNA. This is a very specific region of the DNA double helix. Once bound there will be an increased expression of the activity of the target molecule. This type of expression of activity is slow due to the transcription and translation to the mRNA.

ii. The other mechanism of action for this signaling is the ligand binding to receptors found on the cell membrane. There are 3 general categories of cell –surface receptors. These categories are based upon the mechanism of the signal transduction.

a. The receptor and the ion channel are part of a single multimolecular complex. The best studied of these are the ligand gated ion channels of the GABA and the nicotinic acetylcholine receptors of the muscle cell.

b. Catalytic receptors are known to include an enzyme like activity in their structure. The best understood of this classification is the tyrosine kinase receptor.

c. Receptors that involve the use of a second messenger system. In this system, there is a binding of a substrate to a specific receptor on the extracellular side of the cell membrane and then this binding will initiate a series of reactions that will produce a specific response intracellularly. Often this mechanism results in signal amplification. Prototypic receptors are known to have an extracellular domain and a helix that will cross through the cell membrane 7 times. The best understood of the second messenger systems is the adenylate cyclase system and the calcium / phosphatidylinisitol system.

B. Adenylate Cyclase second messenger system is a system that is membrane bound and it is designed to convert ATP to 3'5' – adenosine monophosphate AKA cAMP or cyclic AMP.

i. In the reception of the substrate to the receptor, there will be either an increase or a decrease in the activity of adenylate cyclase activity. Prototypic receptors are known to have an extracellular domain and a helix that will cross through the cell membrane 7 times. These interact with G proteins to elicit their action. These G proteins are trimeric proteins that are called G proteins due to the guanosine nucleotides that they are bound to. The inactive form of the G proteins are bound to GDP. This trimeric G protein will dissociate into an α subunit with a βγ dimer. The active form is now a GTP bound form of the α subunit. This allows for the amplification of the adenylate cyclase to be effected, allowing many active forms of G proteins to be generated from a sole activated receptor. In this system, there is a specific G protein, G_s that is a stimulatory protein. There is also a specific protein, G_i that inhibits the activity of adenylate cyclase. The activity of this protein complex is responsible for the hydrolysis of GTP just as we see ATPase hydrolyze ATP to ADP + Pi. This will inactivate the G protein.

ii. Activation of cAMP (cyclic AMP) is done with cAMP dependent protein kinases. This is the next step in the second messenger system. Activation of cAMP is also possible through another enzyme, protein kinase A. cAMP will activate the protein kinase as it binds to one of the regulatory subunits. Binding to the subunit will release the catalytic unit. The now activated subunit will catalyze the transfer of phosphate from ATP to specific serine or Threonine residues. The now phosphorylated subunits are able to act on the cell ion channels without the inclusion of another subunit or they may activate or inhibit other enzymes. It is the phosphorylation of the protein kinase A that will activate the enhancer regions of the DNA to increase the expression of the gene. Not all protein kinases will respond to cAMP. In order to hydrolyze the phosphate from the enzyme a second enzyme must act. The enzyme responsible for the removal of the phosphate is protein phosphatase. Further hydrolysis of cAMP is done through the enzyme activity of phosphodiesterase as it cleaves cAMP to 5' AMP. Once the cyclic 3'5' diesterase is cleaved, the remaining 5'AMP is no longer an intracellular signal.

C. Calcium / phosphatidylinositol system of second messengers is well understood. In this system, the receptor mediated signal transduction uses G proteins as the cAMP system does. However, in this system phospholipase C, a membrane bound phosphodiesterase, will be activated and elicit a cascade effect. The activated phospholipase C will direct the cleavage of an internal membrane bound phosphatidylinositol 1,4,5 triphosphate to yield diacylglycerol and 1,4,5 triphosphate. This is important as the two of these will act in concert as a second messenger.

i. The inositol 1,4,5 triphosphate is specific in the manner the inositol will become bound to the endoplasmic reticulum eliciting rapid release of stores of calcium. This is imperative for the formation of calcium calmodulin complexes. Cleavage of inositol 1,4,5 triphosphate will yield subsequent inactive products, inositol 1,4 bisphosphate and 1 inositol phosphate.

ii. The DAG (diacylglycerol) that is yielded in the cleavage by phospholipase C of phosphatidylinositol 1,4,5 triphosphate will activate protein kinase C. This enzyme is dependent on calcium for effective functioning. In this situation, the activity of the DAG will be affected by the increased affinity of protein kinase C for calcium.

D. Nitric Oxide is also a functional second messenger. Nitric oxide functions in the same role as endothelium relaxing factor. The physiologic activity of nitric oxide is for the relaxation of endothelial smooth muscle walls, specifically, the vascular walls. Therefore it is a functional vasodilator. It is derived from the combination of arginine, O_2, NADPH, FMN, FAD, heme, tetrahydrobiopterin as they are acted upon by the enzyme NO synthase. There is substantial evidence that TNF and IL-1 will stimulate the synthesis of this enzyme.

VII. Measurement outcomes for metabolism.

A. In looking at and understanding metabolism, there must be a measurable parameter for the results of the processes of anabolism and catabolism. There are many very reliable cellular experimental parameters that are used but there are tools that are available without looking at the cellular level. These are the Anthropometric measurements of nutritional assessment. The use of the height to weight charts are designed to correlate the balance of under nutrition to over nutrition in pregnant women, children and adolescents.

B. Percentages of desirable body weight (%DBW) or percentages of ideal body weight (%IBW) and percent of usual body weight (%UBW) are al parameters that are utilized when investigating the nutritional status of patients.

C. Head circumference is a measurement that is used in children under 2 years of age to correlate the brain growth and development.

D. Midarmcircumference,Midarmmusclecircumference and flat fold tests are designed to investigate the protein status and the amounts of subcutaneous fat.

Vitamins

I. Vitamins are defined as organic compounds that the human body is unable to synthesize. These compounds are required for normal development and growth. The term 'Vitamin' applies to any compound or closely related group of compounds satisfying the following criterion that it is organic in nature (contains carbon), it cannot be synthesized by body in adequate amount, it must be included in diet so that when absent, results in a deficiency state.

Vitamins are essential for normal growth and health. Vitamins are present in foods in small concentrations, and are not a CHO, saponifiable lipid, amino acid or protein. Vitamins function in the human body as an enzyme precursor, coenzyme, often in association with a mineral cofactor. Vitamins are also destroyed by heat.

A. Humans have two sources of vitamins

i. Foods – the problem is that there is no singe food to supply the total vitamins required for good health.

ii. Synthesized by the bacterial intestinal flora, but again there is no synthesis of the total requirements for good health.

B. Vitamins are classified by solubility. This solubility is either water soluble or fat soluble.

i. Water soluble vitamins are B vitamins, folic acid, niacin, panthothenic acid, biotin, and Vitamin C.

ii. Fat soluble vitamins are Vitamin, A, D, E, K. Humans are only able to synthesize Vitamin D. structurally they appear $-CH_2-C(CH_3)=CH—CH_2-$

C. Cofactors – vitamins are often referred to as coenzymes. This is due to the function of vitamins often being as adjuncts to the function of an enzyme. In many instances, the enzyme requires conjugation with the protein portion of the enzyme, called the apozyme. Often times the enzyme has the cofactor bound to the molecule, called hololzyme. If the cofactor is bound tightly, and does not dissociate, it is called a prosthetic group.

D. Minerals may also serve as cofactors. Most often these are the transition metals (zinc, iron, copper) that fill this role.

E. How do we end up with deficiencies. Most often the vitamin or mineral deficiencies are due to a variety of factors: Inadequate absorption, Inadequate use, Increased requirements, increased excretion and externally induced deficiency.

i. Inadequate absorption may be from biliary obstructions as lack of bile creates a lack of the absorption of the fat soluble vitamins. Ileitis, pernicious anemia and celiac disease may also produce poor absorption of vitamins.

ii. Inadequate use may be due to the lack of a serum transport protein or lack of synthesized end product.

iii. Increased requirements of the vitamin may be from growth, pregnancy, trauma, lactation.
iv. Increased excretion may be from renal dysfunction
v. Externally induced deficiency may come from the loss of microbial flora with antibiotics or other drug therapies.

F. Axerophthol, retinol, Vitamin A: this is an alcohol and aldehyde retinol and retinoic acid. These are isoprenoids that are known to have lipid characteristics.

i. Vitamin A is derived from β carotene, a provitamin. The vitamin A that is derived is required to be absorbed with fat and must be stored within the liver. There is a modification that must be done prior to this storage and transport. Vitamin A must be esterified as retinyl palmitate for it to be utilized in this capacity.
ii. Prior to transport in the blood, there is a need for the ester to be hydrolyzed eliciting the release of the retinol.
iii. Once released, the retinol binding proteins must bind the retinol for transport.
iv. Vitamin A is extremely contraindicated in pregnancy, even in very small amounts. During pregnancy, Vitamin A is considered a teratogen. This is because Vitamin A is involved in the regulation of embryogenesis, fertility, cellular differentiation, and growth.

iv. Retinoic acid is involved in gene expression by its ability to bind to the transcription factor RAR receptor.

v. Retinyl phosphate is known to be involved in the synthesis of glycosaminoglycans.

G. Water soluble vitamins are most often cofactors in enzyme systems.

i. Thiamine – aneurin (B1) - Metabolism in the example of thiamine is the precursor of thiamine pyrophosphate (TPP). This is the unphosphorylated form of thiamine. In order to be converted to the thiamine pyrophosphate, there must be a phosphorylation reaction. This reaction is run in the liver with ATP from the liver as the phosphate donor. This reaction is also run in the muscle, brain and heart. The stores of this are only available for up to 12- 14 days. The individual aspects of the coenzyme are the pyramidine and the thiazole. Each of these components will be synthesized individually. The function of this coenzyme is the transfer of the aldehyde groups.

ii. Deficiencies of this vitamin will produce dry beriberi with chronic deficiency. Symptoms are peripheral neuropathy. Wet beriberi, seen when the deficiency is more severe and chronic with neurological symptoms and tachycardia. Additional pathologies are Wernicke-Korsakoff syndrome, with the main population affected being alcoholics.

iii. Pyridoxine (B6) is unusual in that there are several compounds that can substitute as the B6 itself, (pyridoxamine pyridoxal). The nonphosphorylated form of this supplement is absorbed in the upper GI. Requirements of B vitamins are related to the protein use, and it is increased during pregnancy, periodic growth, lactation and wound healing. Urinary excretion of riboflavin is decreased in positive nitrogen balances.

II. Minerals constitute about 4% of body weight in the normal adult. Minerals are able to provide many varied functions that are essential for life for example they assist in maintaining the osmotic balance, catalytic functions for vitamins and enzyme reactions, structural component of the bony skeleton. Unlike the vitamins discussed previously, the minerals are heat stable. Chelates stabilize otherwise unstable compounds such as Mg-ATP by forming a complex coordination organometallic bond. Metals act as Lewis acid catalysts, especially with transition state metals Zn, Fe, Mn, and Cu which have empty d orbitals that can act as electron sinks. Minerals may function to stabilize a transition state and prevent intramolecular damage such as REDOX reactions with -OH groups by binding Hydrogen from water. Minerals also stabilize tetrahedral or other complex structures via coordinate bonds as with Fe-Oxygen in Hemoglobin. Inorganic minerals play an important role in enzyme reactions.

Minerals facilitate electron transfers, and enable a substrate to undergo changes in redox state without measurable changes in pH of the milieu.

III. Macrominerals (> 0.01% of Body weight, > 100 mg per day) vs. Microminerals (< 0.01% of BW, < 100 mg/day needed)

A. Macrominerals include:
Calcium
Phosphorus
Magnesium
Sodium
Potassium
Chloride

B. Microminerals include:
Iron
Zinc
Copper
Selenium
Iodine
Manganese
Molybdenum
Fluorine

C. Trace and Ultratrace minerals are also required for normal physiologic and biochemical functioning.
Nickel
Silicon
Vanadium
Arsenic
Boron
Cobalt

It is important to understand that many mineral present in the body by 'association', are found in the soils where the plants we consume happen to grow, have neither positive or negative roles. The current scientific literature has demonstrated that many minerals are toxic, even when consumed in small amounts (i.e., lead, mercury). The adverse and pathologic and possibly lethal effect of these minerals has been understood for hundreds of years. Some essentials toxic when consumed in excess (ex: selenium, cobalt, arsenic).

Oxidative Phosphorylation And Mitochondrial Electron Transport Chain

I. Energy must be transferred, produced and used within biologic systems. The description of these processes is the Bioenergenics of the system. The description of this system mathematically, is very similar to thermodynamics. It is important to understand the energy balance in normal physiology, energy in should equal energy out + energy stored. $E_{in} = E_{out} + E_{stored.}$ This is true since majority of the energy used is in the form of heat energy. Bioenergetics is only concerned with the initial and the final energy states of the reactants. To describe this the first term required will be that of free energy.

A. Free energy or changes in free energy (ΔG) will provide the information required to decide if the reaction is going to run or not. Mathematically, $\Delta G = \Delta H - T\Delta S$.

i. The sign of ΔG will predict the direction of the reaction if the temperature and pressure is constant.

ii. If the ΔG is negative, the reaction will be endergonic, a net loss of energy and the reaction will run spontaneously as written.

iii. If the ΔG is positive, the reaction will endergonic, and there must be energy put into the system for the reaction to run. It therefore, will not run spontaneously.

iv. If $\Delta G = 0$, the reaction is in equilibrium.

v. ΔG is dependent on the [reactants] and [products] so the mathematical expression of this will be $\Delta G = \Delta G^\circ + RT \ln [B] / [A]$. The signs for ΔG and ΔG° can be different. This is possible when the reaction with a positive ΔG° can proceed forward and have a negative overall ΔG if the ratio of the products to the reactants [B] / [A] is small enough.

vi. Standard free energy (ΔG°) is equal to the free energy change (ΔG) under standard conditions.

vii. The properties of ΔG° make it additive if the reactions are consecutive in reaction order. This is also true of ΔG as it pertains to the reactions order. Just as the ΔG's are additive, the sums of the reactions will decide if the total reaction order will proceed as spontaneously written or not.

B. In order to determine the direction and the extent of the reaction running, there are two factors that must be dealt with.

i. Enthalpy (ΔH) describes the change of heat between the reactants and the products.

ii. Entropy (ΔS) describes the randomness or the degree of disorder of the reactants and the products. This value does not predict if the reaction will run.

II. the Electron transport chain is also known as the respiratory chain. This biochemical path is one method that the mitochondria utilize to push electrons from variety of substrates to molecular oxygen. Glucose and fatty acids are metabolized to yield the final products of CO_2 and water. It is the intermediates of the path that will produce the energy product of ATP.

A. This pathway is a series of oxidation and reduction reactions.

B. Reduced A + Oxidized B <-> oxidized A + reduced B

C. In this pathway it is possible that many substrates can use a common pathway since many will be oxidized by enzymes using oxidized nicotinamide – adenine dinucleotide (NAD+) or oxidized flavin – adenine dinucleotide (FAD+) as the electron accepting cofactor.

D. In this pathway, the electrons from the reactants are donated to the carriers and as this is done, there is a loss of free energy. The loss of energy is in turn captured and used to produce Adenosine diphosphate and inorganic phosphate. This is the coupling of the oxidative phosphorylation to the electron transport chain.

E. Energy that is not converted to the nucleotide adenosine diphosphate and the inorganic phosphate will be transferred as heat energy.

F. The inner mitochondrial membrane is the location of the complexes of the electron transport chain.

   i. The inner mitochondrial membrane contains a large concentration of membrane proteins. These proteins will constitute the complexes of the electron transport chain.

ii. The membrane is permeable to a variety of ions and small molecules.

iii. The chain is highly organized as complexes 1, 2, 3, 4, & 5. Complexes 1-4 are the actual components of the electron transport chain. Complex 5 is also known as the ATP synthase complex. These complexes will be expanded on later in this chapter.

iv. All of the complexes have the ability to accept electrons from the previous carrier via the electron donors and then pass those electrons along to the next complex using the same carrier system.

v. Molecular oxygen and protons will combine to form water and as such makes this the largest consumer of molecular oxygen.

Fig 13

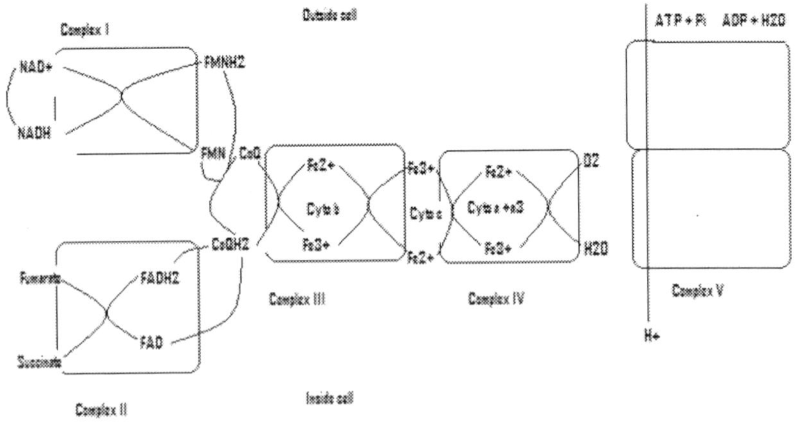

Electron Transport Chain Inner mitochondrial membrane

G. The specific reactions of the chain are proteins with one exception, Coenzyme Q. Cofactors for these reactions are the cytochromes, the coppers, iron, and the porphyrin rings.

i. The reactions require the formation of NADH from NAD+ as it is reduced by the dehydrogenases. In this reaction, there is a transfer of the hydride ion to NAD+ for the formation of the NADH. There is a resultant proton that is unbound.

ii. NADH dehydrogenase that is bound to the mitochondrial inner membrane is the next stop for the proton and the hydride ion. The 2 hydrogen atoms are transferred to flavin mononucleotide making this $FMNH_2$ and $FADH_2$.

H. Cytochromes are the remaining acceptors of the electrons within the chain. The cytochromes are derived from the porphyrin rings mentioned previously. The electrons move through the complement of cytochromes from co enzyme Q to cytochromes b, c, and $a + a_{3.}$ It is this last cytochrome that is the only cytochrome with the ability to react with molecular oxygen directly. This is the conduit for the combination of the components of the chain to be placed together so as to produce water. For this reaction with cytochrome $a + a_3$ to run correctly, there must be bound copper.

I. Free energy release is the goal of the electrons as they are moved along the electron transport chain. The advantage of this system is the ability of the system to use differing electron carriers. Electrons can be transferred as hydride ions, :H- to NAD+, as electrons ·e- to the cytochromes, or as hydrogen atoms to FAD, FMN or coenzyme Q. This makes the electron transport chain a very versatile transfer mechanism and effective transport mechanism.

i. The ΔG° is intimately related to the $\Delta E_\circ$ through the formula $\Delta G^\circ = n\ F\ \Delta E_o$

III. Oxidative Phosphorylation is the primary source of energy in the aerobic cell. Free energy of this pathway is released as the electrons are shuttled along the electron transport system and it is coupled with the formation of high energy phosphate compound adenosine triphosphate from adenosine diphosphate and inorganic phosphate.

In damaged mitochondria, the process described above is uncoupled from the oxidative phosphorylation. This uncoupling allows release of free energy release but not as ATP generation but rather as heat energy.

IV. Components of the Electron Transport Chain

A) Outer membrane - permeable to small molecules

B) Intermembranous space

C) Inner membrane – is very highly selective

i) Specific transport system for: ATP, ADP, Phosphate

ii) Pyruvate, succinate, alpha keto –gluterate, malate, citrate

iii) Cytidine & Guanosine phosphate, (both tri & diphosphates)

D) The enzymes of the Electron Transport chain is found within the inner membrane along with the enzymes of the oxidative phosphorylation

E) The matrix abuts the inner membrane with the enzymes of the citric acid cycle (TCA), enzymes associated with the beta –oxidation of fatty acids.

F) Sources of electrons are the following: Reduced nicotinamide- adenine dinucleotide (NADH) that is derived from NAD+ dehydrogenases, isocitrate, alpha – ketogluterate and malate dehydrogenases of the TCA, Pyruvate dehydrogenase, L-3-Hydroxylacyl coenzyme A (CoA) dehydrogenase (fatty acid oxidation)

G) Sources of Flavin – adenine dinucleotide ($FADH_2$) include: succinate dehydrogenase of the TCA, FAD –linked dehydrogenase of the alpha – glycerophosphate shuttle, Acyl CoA dehydrogenase of fatty acid oxidation.

V. The Complexes associated with the Electron transport chain

A) Complex 1 – point of entry for electrons of NADH. The enzymes that catalyze this are NADH – coenzyme Q reductase or NADH dehydrogenase

i) Prosthetic Groups – flavin mononucleotide (FMN) / Iron - sulfur centers (Fe-S)

ii) Electron acceptor – coenzyme Q (ubquionone)

iii) Pathway: electrons from NADH -> FAD -> Fe-S centers -> Co Q

iv) Inhibitors – Rotenone (insecticide), Amobarbitol & secobarbitol, Piericidin A (antibiotic)

B) Complex 2 – point of entry for electrons from succinate. The enzyme complexes that catalyze this are: succinate – Co Q reductase. This complex includes succinate dehydrogenase. This is the same enzyme that participates in the TCA cycle.

i.) Prosthetic Groups – FAD, Fe – S centers, Heme

ii.) Electron acceptors – Co Q

iii.) Pathway: Succinate -> FAD -> Fe-s Centers -> Heme -> Co Q

C) Co Q – this is a very lipid soluble molecule that is found in the mitochondrial membrane.. It is found to exist in all living cells. The function of this complex is to accept the electrons from complex 1 and complex 2 and donates them to complex 3.

D) Complex 3 – acceptor for electrons from Co Q

i.) Prosthetic groups – Heme, Fe-S centers

ii) Electron acceptors – cytochrome c

iii) Pathway: CoQ -> Heme -> Fe – S centers -> Cytochrome C_1 -> cytochrome c

iv) Inhibitors: Antimycin A (antibiotic)

E) Cytochrome C – this is a soluble protein that binds to the membrane in order to fulfill its role in the electron transfer. It mediates the transfer of electrons from complex 3 to complex 4.

i.) Prosthetic Groups – heme

F) Complex 4 – this complex is the electron acceptor for cytochrome c. The enzyme complex that catalyzes the reaction is cytochrome c oxidase.

i.) Diseases associated to this are: Leigh Disease, Alper's Disease, Myoclonus Epilepsy.

ii.) Prosthetic Groups – Copper, heme(cytochrome a and cytochrome a_3)

iii.) Electron Acceptor – molecular Oxygen

iv.) Pathway – cytochrome c -> Cu – cytochrome (a types) -> molecular oxygen

v.) Product – water

vi.) Inhibitors – CO, Hydrogen sulfide (H_2S), Azide, Cyanide (CN-)

VI. Formation of Adenosine Triphosphate

A) The energy of this system is 52 kcal for every pair of electrons that pass along the chain. This 52 kcal produces 3 moles of ATP via the phosphorylation of ADP.

Fig 14

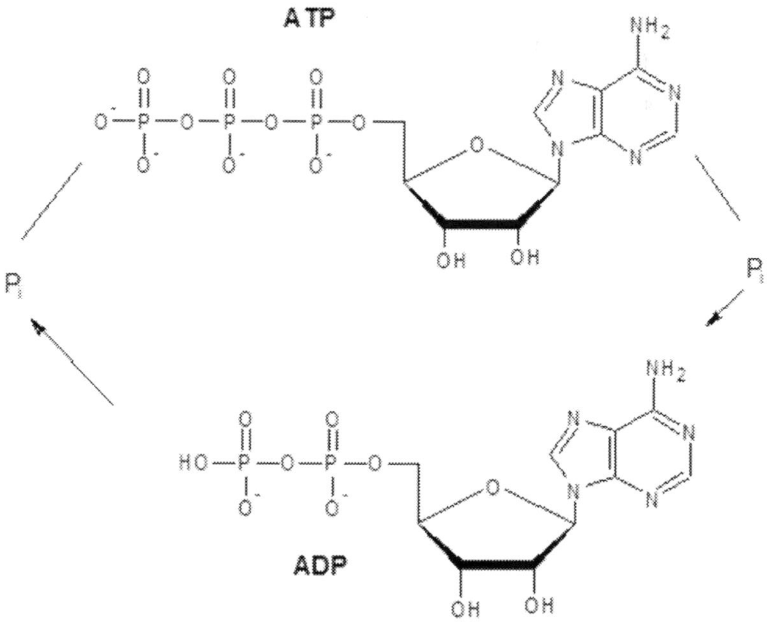

ATP

ADP

i. The ΔG° of ATP is -7300 cal/mol for each of the terminal phosphate groups.

B) Coupling sites for the synthesis of ATP are Complex 1, Complex 3 and Complex 4. These complexes are sufficient to generate 1 mole of ATP.

i) Electrons that enter the chain from NADH will generate 3 moles of ATP

ii) Electrons that enter the chain from $FADH_2$ generate only 2 moles of ATP. This reduction of ATP production is because Complex 1 is bypassed.

C) Chemiosmotic coupling states that the electrochemical gradient of protons (H+) across the inner mitochondrial membranes produce the energy required for the manufacture of ATP. This electrochemical gradient means that there must be a pump for this exchange of protons (H+) to occur. In this model, the electron movement along the chain will elicit the pumping of these protons into the intermembranous space from the matrix and it is this movement that can either generate ATP or heat. The pump is designed to pump the protons in such a manner so that there will be an end product of ATP. The theory of this is that the transfer from the inner mitochondrial membrane to the cytosolic side is the flow of protons that will yield the high energy phosphate compound. The protons are able to gain entry back into the mitochondrial matrix through the channel designated as ATP synthase. It is the passage of the protons through this channel that is the point of phosphate generation. While the synthesis of ATP is being initiated, there is concurrent dissipation of the pH gradient.

D) ATP Synthase is the actual complex that is associated with the synthesis of ATP. This complex is also known as the H+ - ATPase complex. The structure of this complex is extremely unique. It consists of two subunits,

i) the F_0 subunit spans the membrane and it is further subdivided into four polypeptide units. These polypeptide units form a channel that the protons (H+) use to pass through the membrane.

ii) the F_1 subunit is bound to the F_0 subunit within the matrix of the mitochondrial membrane. This subunit is unique in its structure as well since the formation of this subunit is five polypeptides. The function of this subunit is also very different from the other side in that this subunit is designated to be a catalytic site of the synthesis of ATP.

iii. This is the unit that is the emphasis of the entire coupling of the electron transport chain and the oxidative phosphorylation. This is where the ATP is generated through the gradient that is established by the complexes of the system. This is also the point of the chemiosmotic theory as the electrons that have been transferred to the cytosol will re-enter the cell via this ATP synthase and change the pH as well as the electrical gradients of the cell membrane by having a greater concentration of protons on the outside of the cell. This is all concurrent with generation of the ATP that was the original goal of the system in the first step.

iv. Inhibitors – Oligomycin / Cicyclohexylcarbodiimide (DCCD)

v. Additional coupling enzymes include the following complexes: Calcium – ATPase, sodium – potassium – ATPase.

VII. Integration of the Oxidative portion and the Respiration rate.

A) The ability of the body to control ATP synthesis within the mitochondria is controlled by ADP concentration. This regulatory control is achieved via a negative feedback loop.

B) There are factors that regulate this respiratory control and they include the following:

i) limited availability of ADP and/or substrates (electrons)

ii) limited availability of substrates

iii) the capacity of the complexes of the electron transport chain themselves.

iv) ADP availability

v) Oxygen availability

C) There is a mathematical description of the amount of moles of ATP that are generated and this is the P:O ratio. It defines how many moles of ATP are generated from ADP per gram atom of oxygen for a given substrate.

i) Substrates that donate to NAD+ yield a P:O ratio of 3:1

ii) Substrates that donate electrons to FAD+ yield a P:O ratio of 2:1

D) What will uncouple the oxidative phosphorylation.

i) 2,4 Dinitrophenol

ii) Dicumarol – anticoagulant

iii) Chlorocarbonylcyanide (CCCP)

iv) Bilirubin

VIII. Superoxides are a toxic byproduct of the metabolism of molecular oxygen. In this form of metabolism, the molecular oxygen becomes the anionic form of O_2^- (a free radical). This form is very unstable and very reactive.

A) Formation of this reactive species is within the mitochondria and using the reducing agent FADH and reduced CoQ.

B) The toxic effects of these superoxide free radicals is countermanded by the action of Superoxide dismutases.

C) Hydrogen peroxide is also produced in the clearance of the superoxide radicals. The ability of the body to negate the effects of hydrogen peroxide is accomplished by the enzyme catalase.

D) Within the Erythrocyte there is a like enzyme working to protect the cell from the reduced glutathione that is called glutathione reductase.

E) Vitamin C and vitamin E are also considered to be superoxide scavengers.

VIV. Oxygen Toxicity – high levels of oxygen can be toxic to infants due to the production of the aforementioned superoxide free radicals. This toxicity is even more problematic in premature infants due to the limited ability of the premature infant to produce the superoxide dismutase. The problem most often associated with this oxygen toxicity is blindness.

Proteolysis

I. Proteolysis is the catalytic breakdown of protein into the free amino acids and then further into the constituents components of CO_2 (carbon dioxide) and ammonia. In terms of the amounts of proteins available, the term used is "Protein Pool" This pool is made up of the total body proteins and it is combined with the proteins absorbed from the gastrointestinal tract in the form of amino acids. Outputs of this pool will provide amino acids for the synthesis of proteins and synthesis of special molecules.

A. The concentration of amino acids is greater in the intracellular compartment than it is in the extracellular compartment.

i. Gradient is maintained by the requirement of energy expenditure required for active transport of the amino acids across the cell membrane.

ii. This gradient is a variable figure as it varies depending on the amino acid described. The greatest disparity is seen in the amounts of glutamate and glutamine.

iii. Extracellularly, the protein content is composed of endogenous proteins of the skeletal tissue and supportive tissues. Histologically, the greatest amount of this protein is collagen. The other small but very important component is in the plasma proteins, like albumin. Albumin is a protein that is vital in its ability to carry other molecules and it has a very short half life, of 20 days. This means that there must be available amino acids to replenish this protein constantly.

iv. Intracellularly, the digestive proteins utilized as enzymes are the bulk of these proteins. Examples of these proteins are the Chymotrypsin, Elastase, Pepsin, Trypsin, Aminopeptidase and Carboxypeptidases.

B. The total amount of protein in the human body is around 100g. 50g of this total is derived from glutamine and glutamate with an additional 10g coming from the amino acids with the designation "essential " amino acids.

i. Essential amino acids are those amino acids that are not able to be synthesized within the body.

ii. This inability to synthesize specific amino acids requires dietary supplementation. The common acronym for the actual amino acids designated as essential is :PVT TIM HALL.

Phenylalanine Valine Threonine Tryptophan Isoleucine

Methionine Histidine lysine Leucine

iii. Inability to obtain these amino acids creates a pathologic situation. The pathologies will range from neurological defects to hematological malformations.

C. Amino acid inputs are from the dietary proteins taken in. This figure is between 60g – 100g in most individuals. There is conflicting reports of dietary requirements in elite athletes.

i. Within the gastrointestinal tract there is a larger amount of protein turnover seen just in the "sloughing off" of the intestinal mucosal cells as they are degraded.
ii. The actual digestion of the dietary proteins is accomplished through the proteolysis via gastric juices and pancreatic juices.
iii. Hydrochloric acid has a pH of 2, making it a very powerful acid. The action of this acid is to denature the proteins and provide an acidic environment for the activity of pepsin.
iv. As mentioned above, the activity of pepsin requires a strongly acidic environment for optimal activity. It is secreted as the proenzyme pepsinogen that requires the cleavage of a 44 N terminal amino acid chain or by pepsin itself. This is an example of an enzyme that can undergo autoactivation and autocatalysis. The activity of this enzyme will be the liberation of majority of the amino acid residues. These liberated amino acid residues (peptide chains) will stimulate additional enzyme activity, specifically, cholecystokinin in the duodenum.
v. Pancreatic juice is rich in carboxypeptidases and endopeptidases. These enzymes are also secreted as proenzyme from the pancreas. Note that it is the pancreatic acinar cells that are responsible for the release of these non activated enzymes. There are

specialized peptides for this stimulation to be done, secretin and cholecystokinin – pancreozymin. For the conversion of the inactive preproenzyme to fully activated enzyme, the action of a enteropeptidase is required. The enteropeptidase will cleave the six amino acid residues required for the conversion of the trypsinogen to Trypsin. This conversion creates a feed forward loop of stimulation for the conversion of trypsinogen to Trypsin. The activity of Trypsin also stimulates the conversion of the proenzymes Chymotrypsin and Elastase as well as the exopeptidases carboxypeptidases A & B.

vi. Aminopeptidases in conjunction with dipeptidases are going to continue the cleavage and degradation of the proteins into the amino acid components.

D. Proteolysis of endogenous proteins is a process that is variable as some proteins are degraded within hours and other are active for days while structural proteins remain viable for very extended times. There are many factors that will affect the rate of degradation of these proteins such as lysosomal activation, activity of glucocortocoids and activity of thyroid hormones. These peptide hormones will increase the turnover rates of proteins while insulin is a peptide hormone that slows the catabolic process down. It is actually considered to be an anabolic hormone.

E. The proteins from the amino acid pool provide the building blocks for many proteins synthesized in the body. The catabolism of the amino acids from this amino acid pool will yield urea and CO_2. These process are a continuous drain on the amino acid pool.

II. Nitrogen Balance defines the equilibrium between the intake and excretion of the nitrogenous bases that make up the amino acids. The initial aspect of catabolism removes the α amino group through a transamination reaction or oxidative deamination. The product of this reaction is ammonia and the corresponding α ketoacids. These α ketoacids are going to donate the carbons for conversion to intermediates of the metabolic pathways that will yield end products of CO_2, glucose , fatty acids or H_2O.

A. In the urine we excrete 90% of our nitrogen as urea. Additional loss is from the fluid loss associated with sweating and defecation. There is no storage mechanism of amino acids like there is for fatty acids. In the situation of excessive intake of proteins, the metabolism of the nitrogen to urea will yield an end product of ammonia. This can be toxic if the excess is prolonged.

i. There are specific mechanisms for the digestion of the dietary protein intake.

ii. Gastric secretions include hydrochloric acid with the proenzyme pepsinogen.

iii. Pepsinogen will become activated to pepsin after secretion of the zymogen form from the serous cell of the fundus of the stomach. The HCl is the activating factor for the pepsinogen to be converted to pepsin. There is also another mechanism for the activation of the zymogen, autocatalytic activation by previously active pepsin.

iv. The metabolism continues as peristalsis moves the intake into the duodenum. The large macromolecules that have been acted upon by the enzyme pepsin will be further cleaved by additional pancreatic enzymes. Another pancreatic enzyme, Trypsin will act at the carbonyl only is the carbonyl is from arginine or lysine. Again this enzyme is secreted as a zymogen and must be autocatalytically activated. The release of the digestive enzymes are regulated by two hormones, cholecystokinin and secretin.

v. Enteropeptidase that is synthesized on the intestinal mucosal cells will convert the trypsinogen to Trypsin through the cleavage of the hexapeptide at the N terminus of the proenzyme.

B. If there is a net gain in the amount of nitrogen produced versus that lost there is a "Positive Nitrogen Balance" and this is deemed to be an anabolic process.

C. Conversely, if there is a net loss of nitrogen, then this is termed a "Negative Nitrogen Balance" and this is a state of catabolic activity. In this situation, should the total loss equal 1/3 of the total body protein it may prove to be fatal.

D. The location of the amino acids that are in the extracellular compartment is significantly less than those that are found intercellularly. This balance is maintained through a variety of mechanisms. There are 7 transport systems that are well understood for the transport of thc amino acids. The kidneys are responsible for the absorption of cystcine, ornithine, arginine and lysine.

E. Tansamination will yield α keto acid and glutamate.

F. Aminotransferases are specific for the amino donor as the acceptor is most often α ketogluterate.

   i. Alanine aminotransferase (ALT) or glutamate pyruvate transaminase will transfer the amino of Alanine to the α ketogluterate. This reaction will yield pyruvate and glutamate.

   ii. Aspartate aminotransferase (AST) or glutamate oxaloacetate transaminase (GOT) will transfer amino group from the glutamate to Oxaloacetate yielding aspartate.

Fig 15

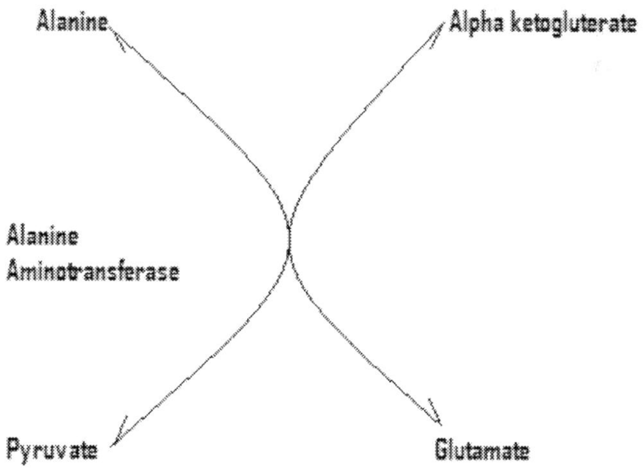

Alanine
Alpha ketogluterate
Alanine Aminotransferase
Pyruvate
Glutamate

Fig 16

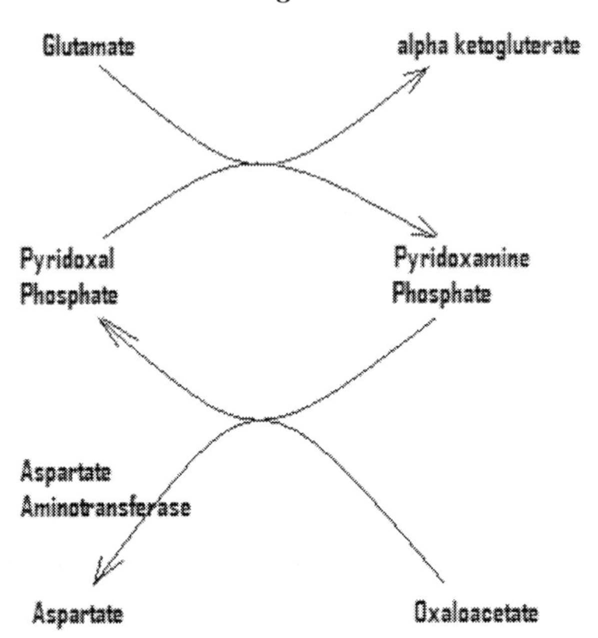
Glutamate
alpha ketogluterate
Pyridoxal
Phosphate
Pyridoxamine
Phosphate
Aspartate
Aminotransferase
Aspartate
Oxaloacetate

Carbohydrates

I. These are carbon molecules that are hydrated. Carbohydrates (CHOs)yieldapproximately4kcal/gram. Thesecarbohydrates can be simple sugars (sucrose, dextrose) or possibly seen as complex carbohydrates (starches). Carbohydrates are also seen as fiber: (not digested in humans, therefore no kcals for human digestion however there is about 1 kcal/g via colonic microorganisms)

Because of this configuration, these molecules can be polyhydroxylated ketones, polyhydroxylated, aldehydes, or other compounds that can be hydrolyzed into these compounds. These compounds have a variety of functions that range from energy, structural components, provide the backbones of nucleotide formation, act as metabolic intermediates and form Synovial fluid.

Fig 17

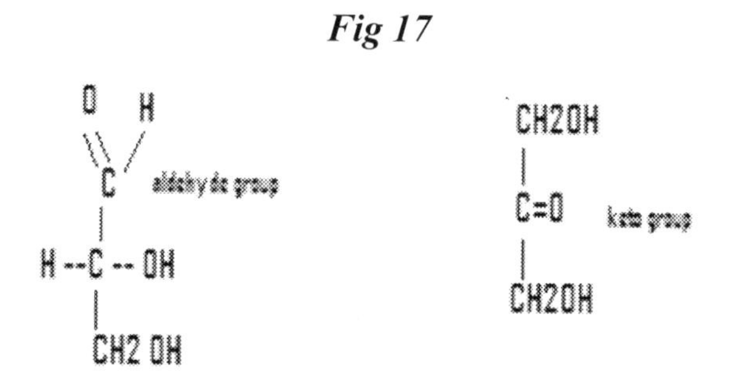

A. Carbohydrates provide the major share of the cellular energy as sugar constituents.

B. Carbohydrates are known to provide structure in cell walls in both plant and animal cells.

C. Carbohydrates also act as metabolic intermediates.

D. Carbohydrates also make up the bulk of the backbone of the nucleotides in DNA & RNA.

E. Carbohydrates act in the cell recognition and cell communication via the receptor.

F. Carbohydrates serve in the immune system in recognition of the receptor or as a signal for the receptor.

G. Carbohydrates with a free carbonyl group are known to have the suffix "-ose".

II. Classification of Carbohydrates

A. Aldoses – these are monosaccharides with an aldehyde group (CHO) as the reactive group.

B. Ketoses – these are monosaccharides with a ketones (C=O) as the reactive group

C. Monosaccharides – these molecules can be simple sugars (glucose and fructose) or they can be linked to other molecules to form complex compounds as in glycosidic linkages.

i. Disaccharides – sucrose & maltose are able to be hydrolyzed to two monosaccharides

ii. Oligosaccharides – are polymers of two to ten monosaccharides. (blood group antigens)

iii. Polysaccharides – are polymers of more than ten monosaccharide units (starch & Cellulose)

iv. If the oxygen of the carbonyl group of the anomeric carbon is not bound to another molecule, the sugar is classified as a reducing sugar. It is only the oxygen on the anomeric carbon that is responsible for the distinction of that sugar being a reducing sugar or not.

v. In naming the carbohydrate, it is important to remember that the aglycone (the non carbohydrate portion of the molecule) is not responsible for this description.

D. Carbon Numbering - trioses (3 carbons), Tetrose (4 carbons), Pentose (5carbons), Hexose (6 carbons). The carbons are numbered sequentially and the aldehyde or ketone is found on the carbon with the lowest number possible. Numbering begins with the end that is found to contain the carbonyl carbon.

i. Reactive group nomenclature is often used to describe the compound in conjunction with the monosaccharide nomenclature. Glucose is an aldohexose or a 6 carbon monosaccharide that has an aldehyde group.

ii. Glucose is the most common monosaccharide that is consumed by humans.

E. Metabolism of monosaccharides and disaccharides is important since the dietary intake is made up of approximately 20% fructose that is derived from sucrose.

i. Sucrose will yield fructose and glucose when metabolized.

Fig 18

Glycogen
Galactose
UDP Glucose
Galactose 1 Phosphate
UDP Galactose
Glucose 1 phosphate
Glucose 6 hosphate
Glucose

Fig 19

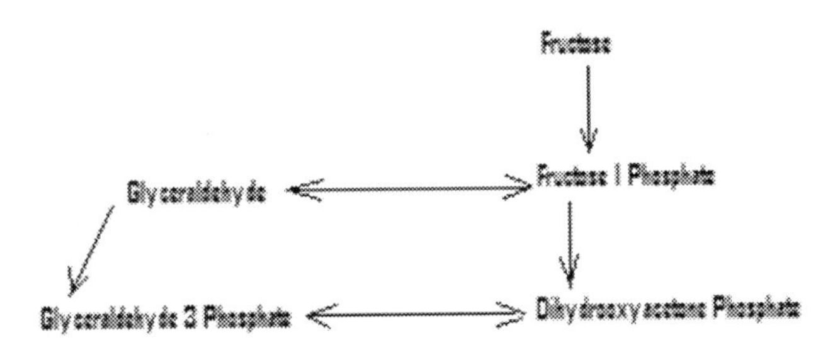

III. Carbohydrate Structure – the structure of carbohydrates are seen to be Open chain or Cyclic forms.

A. In open chain forms there is an asymmetric carbon. Remember that an asymmetric carbon is one that is bonded to four different atoms or groups. In sugars like glucose, the C2 and C5 are asymmetric with this asymmetry being the reason that there is optical activity of the enantiomers. Enantiomers are the mirror images of one another. This concept is demonstrated in the simplest carbohydrates, the monosaccharide trioses (glyceraldehyde). This configuration produces the L and the D optical configurations. In humans, almost all sugars are of the D configuration. As with most biochemical "laws" there are exceptions to this generality and in these sugars, the exception is in L-fructose and L-iduronic acid.

B. Isomers are compounds that have the same chemical formula. Examples of this are in sugars with the chemical formula of $C_6H_{12}O_{6,}$ fructose, glucose, mannose and galactose.

C. Epimers are isomers that differ at one carbon only.

D. Enantiomers are isomers that are also mirror images of each other. These two isomers have identical chemical properties.

E. Hemiacetals can be found in linear form or in cyclic forms. Intermolecular hemiacetals are very unstable and linear compounds that are formed from the reaction of an alcohol and an aldehyde. Cyclic hemiacetals are formed by the same type of chemical reactions.

F. Cyclic Forms are the more common form of the sugars in solutions and also in solids.

   a. In this type of molecule, the ring formations are five or six member rings. The constituents are four or five carbon atoms with one oxygen atom.

   b. Ribose, a pentose sugar will form a five member ring that is called furanose.

   c. Glucose, a hexose sugar will form a six member ring that is a pyranose.

G. Anomeric carbons are the newly formed asymmetric carbons at C1 that are formed by the cyclization that occurs at the carbon bound to the oxygen in the hemiacetal formation.

a. If the hydroxyl on this anomeric carbon is below the ring it is called an alpha position
b. If the hydroxyl on this anomeric carbon is above the ring it is called a Beta position
c. The process that these sugars are able to change position in solution is termed the mutarotation.
d. It is the state of the oxygen within the anomeric carbon that will dictate if the sugar is designated as a reducing sugar or not. The sugar is a reducing sugar if the oxygen on the carbonyl group is not attached to any other structure.
e. The anomeric carbon is able to be oxidized.
f. Oxidation of the $–CH_2OH$ group at carbon 6 will yield a "uronic acid"
g. Reduction of the carbonyl carbon will yield a polyol. The reduction of the hydroxyl group will elicit deoxysugars.

IV. Glycosidic Linkages

A. Hemiacetals that react with alcohols are called Acetals.
B. Sugars can react with alcohols to form acetals called glycosides.
C. If the sugar is glucose, the product of that reaction is glucoside, if the sugar is fructose, then it will form a fructoside and so on.
D. If the side chain (R) is another sugar, then the product is a disaccharide.

E. If the side chain (R) is a disaccharide then the product is a trisaccharide

F. Glycosidic linkages are named by reading from left to right, so sucrose is α – 1,2 glycosidic linkage. Sucrose is an unusual sugar in that the oxygen bond between the glucose and the fructose is between the anomeric carbons, thus there is no free aldehyde or ketone in the sucrose. This is important since sucrose is not a reducing sugar.

G. Maltose has an unattached anomeric carbon that can achieve either an alpha or beta configuration. In this sugar, the configuration can also be either alpha or beta but the linkage is an α 1-4 glycosidic linkage, making maltose a reducing sugar.

H. Polysaccharides are long chains of glycosidic bonds.

a. Starch is the storage forms of the sugar glucose in plants and it is made from the combination of amylase and amylopectin.

b. Glycogen is the storage form of carbohydrates in animals. It is found abundantly in the liver and muscle. It is highly branched form of amylopectin with a α 1-6 branching point every 8 – 10 glucose residues.

V. Carbohydrate Derivatives

A. Phosphorylation is the first step in the metabolism of sugars.

B. Phosphoric acid esters of monosaccharides are the product of the reaction of phosphoric acid with the hydroxyl of a sugar.

C. Amino Sugars have the amino group in place of the hydroxyl. It is possible to have acetylamino group as the replacement. Examples of these amino sugars are glucosamine or galactoseamine (the polysaccharide of cartilage, chondroitin sulfate.

D. Sugar acids are produced from the oxidation of the aldehyde carbon or the hydroxyl carbon or even both.

   a. Ascorbic acid is a sugar acid
   b. Glucuronic acid is another sugar acid and it is involved in the metabolism of bilirubin

E. Deoxy sugars are sugars that contain a hydrogen atom in exchange for the hydroxyl group.

F. Sugar alcohols are also called polyhydroxyl alcohols. These can be formed from the reduction of aldoses and ketoses at the carbonyl carbon.

   a. Reduction of aldoses: glucose produces sorbitol
   b. reduction of mannose produces mannitol
   c. reduction of galactose produces dulcitol

VI. Glycoproteins function in the role of structural support as the components of cell walls and membranes. They also function in the formation of lubricants (mucus), recognition sites, hormones (HCG, thyrotropin), and immunologic participants (Immunoglobulins, complement, interferon)

A. Protein – carbohydrate linkages

   a. O –linked glycoproteins: sugars attached by the hydroxyl of serine or threonine
   b. N – Linked glycoproteins: sugars attached by the amide of the asparginine residue. This classification

has three subclassifications: High mannose, hybrid, Complex. The common factor is that all the major types have a pentasaccharide core that is linked to asparginine via N acetylglucosamine.

B. Glycosaminoglycans are the polysaccharide part of proteoglycans. The structure is unique in that it is made of repeating disaccharide units of glucosamine or galactoseamine are present

a. Heparin is a glycosaminoglycan as well as an anticoagulant. This is found in high concentrations in the mast cell.

b. Proteoglycans or mucopolysaccharides are known for the characteristic bottle brush configuration. The glycosaminoglycans that are bound to the proteoglycan are: Chondroitin sulfate, Dermatan sulfate, Heparan sulfate, Keratan sulfate, Hyaluronic acid.

c. Proteoglycans are the basis of the "ground substance" that is the basis of connective tissue.

VII. Blood Group Antigens are made of the oligosaccharides that are bound to the proteins, lipids, and membrane surfaces

A. Blood types are derived from the differing single sugar moiety on a common oligosaccharide core.

B. Typing:

a. Type O – makes antibodies against A & B. Can take blood only from another O but they are universal donors.

b. Type A – makes antibodies to B only. Able to receive from O & A. Donate to A & AB

c. Type B – makes antibodies to A only. Able to receive from O & B. Donate to B & AB

d. Type AB – makes no antibodies. Universal recipients. Donate only to AB.

VIII. Metabolism of Carbohydrates – carbohydrates are of two distinctions. They are either a ketone or aldehyde compound with an attached hydroxyl groups. The distinguishing aspect is in the number of the attached hydroxyl groups. Carbohydrates are important as sugars in the body as ribose sugars and deoxyribose sugars. They are important in the structural aspects of the cell as evidenced in the linkage to proteins and lipids. It is this capacity that the carbohydrate is seen to act as a signaling portion of the cell membrane in the form of recognition by circulating cells and pathogens. Carbohydrates also are considered as the energy deposits of the cell as they are the storage form in many situations in the body and they are the metabolic intermediates in many metabolic pathways.

A. The most basic and perhaps the most simple form of the carbohydrate is the monosaccharide. This form of a carbohydrate is known to consist of either an aldehyde or a ketone group with 2 or more bound hydroxyl groups $(CH_2O)_n$. In this scenario, the most basic form of the monosaccharide is called a triose where there are three hydroxyl groups attached.

a. In this category of monosaccharides there are two very well described examples, the aldose sugar being glyceraldehyde and the ketose sugar being the dihydroxyacetate.

b. If the aldose has 4 carbons = tetrose
c. If the aldose has 5 carbons = pentose
d. If the aldose has 6 carbons = hexose. The most commonly seen of these are the glucose and fructose compounds.
e. If the aldose has 7 carbons = heptose.

B. It is important to describe a structural configuration at this time to better understand the carbohydrate sugars mentioned above. There is a designation assigned to each sugar of "L" or "D". This indicates the configuration of the asymmetric carbon that is the furthest away from the aldehyde or the ketone group. This is important in describing these sugars with multiple asymmetric carbons.

a. The literature describes a molecule with multiple asymmetric centers and no plane of symmetry as having 2^n sterioisometric forms. Therefore in the original description using glyceraldehyde, an aldotriose, n = 1. This means that there are 2 mirror images of the molecule (enantiomers).
b. If the molecule has additional carbons and they present with the like structure at three positions and differ at the 4th position in such a manner as to be opposites they are then called diasteroisomers.
c. Sugar molecules that differ at only a single asymmetric center are epimers of each other.
d. Ketose sugars are much more basic in design. Dihydroxyacetone as mentioned above is the most basic of the ketose sugars since it is optically inactive. Strangely, the ketose sugars are seen to

contain one fewer asymmetric center than their aldose counterparts with the same number of carbons.

C. Structurally, the sugars described will not remain in the open chain positions in solution, rather they will close ranks and utilize a ring configuration.

   a. In the case of the C1 aldehyde in an open chain, it is understood that this C1 position will react with the C5 carbon position to elicit an hemiacetyl structure. This reaction will continue and close the ring to the pyranose 6 carbon ring structure.

   b. In the case of ketoses, there is a very like reaction that will form a hemiketal when reacted with an alcohol. In this situation the C2 carbon of the open chain will react with the C5 hydroxyl and the end product will be furanose because of the 5 component ring that is formed. (add diagram p467 stryer)

D. The final structural aspect of the monosaccharides are the designations of α and β assignments to the structure. If the hydroxyl bound to C1 is found to be below the plane of the ring it is termed an α sugar. If the hydroxyl is found to lie above the ring then it is called the β form.

E. In sugars they are able to link together in a variety of manners. In the O glycosidic bond, there is a bond that forms between the C1 of glucose and the oxygen atom in the methanol that is called a glycosidic bond. Through these glycosidic bonds, the sugar is able to produce disaccharides and polysaccharides. It is also possible to produce glycosidic bonds through the bonding of the nitrogen atom of the amine to the anomeric carbon through a N-glycosidic bond.

F. Sugars are also able to be phosphorylated. In this reaction, the phosphorylation is a mechanism for the additional formation of the N and O linked glycosidic bonds. A secondary aspect of this reaction is that it is one method for making sugars anionic.

G. Metabolic degradation of carbohydrates is initiated in the mouth. This initial metabolism of carbohydrates is accompoilished via mastication, the actual physical destruction of the macromolecule by the teeth. This mechanical disruption is responsible for the increased surface area that the subsequent enzymatic activity is able to act upon.

   a. The enzyme that is responsible for this initial breakdown is salivary α amylase.
   b. The activity of he salivary α amylase is to break some of the α-(1→4)bonds of the dietary starches that we eat. The activity of this enzyme is only seen in the mouth and the esophagus as the lower pH of the stomach will inactivate this enzyme.
   c. The additional aspects of dictary carbohydrates will be the α-(1→6) bonds that are associated with the glycogen and amylopectin.
   d. The absorption of the byproducts of these metabolic processes will be seen in the jejunum mucosal lining and that of the duodenum.

H. In the accounting of the energy production sources of the human body glucose, fructose via sucrose and galactose are among the most important sugars.

I. In the metabolic pathways of humans, the fructose we consume is brought into the metabolic pathways via an enzyme called hexokinase or by fructokinase. In order for this entry to be allowed, the fructose must first be phosphorylated. It is well published that hexokinase has activity in all cells and that this enzyme has a greater Km for fructose. Fructokinase is seen in the kidney and the small intestine and the liver. In either situation, the product of the reaction is the conversion of fructose to fructose 1 phosphate.

   a. Fructose metabolism is very rapid. This is due to the path for the fructose need not be acted on by phosphofructokinase.
   b. Fructose 1 phosphate is able to be metabolized directly to D-glyceraldehyde or dihydroxyacetone phosphate.

J. Galactose is derived through the cleavage of milk products and the resultant lactose (galatosyl β-1,4 glucose) by the enzyme lactase (β-galactosidase). This is like the fructose in that it must first be phosphorylated prior to being metabolized. This phosphorylation is done using galactokinase and produces galactose 1 phosphate. It is now that the pathway becomes divergent from the previous examples of metabolism of monosaccharides. The galactose must be converted into UDP galactose using the enzyme glucose 1 phosphate-galactose 1 phosphate uridyltransferase.

K. Metabolism of Glycogen that is found within the skeletal muscle, the liver and almost every other cell in very minute amounts. In humans the total glycogen content is approximately 400g within the resting muscle and approximately 100g within the liver. The structure of glycogen has been well studied. It is a branched structure of α D glucose. The structure is unique in that it contains two different glycosidic bonds within the same molecule. The primary glycosidic bond is an α –D glycosidic homopolysaccharide bond. The other glycosidic bond is seen at very specific positions and it is the α 1,6 glycosidic bond. This α 1,6 glycosidic bond is found to be located at every 8 – 10 residues where the branching is located.

a. Glycogen synthesis is from α glucose molecules that are bound. This is a endergonic reaction and will require either adenosine triphosphate or uridine triphosphate.
b. The synthesis is done using only the α –D-glucose molecule that is attached to the uridine diphosphate. The enzyme that is responsible for this initial synthesis is UDP-glucose pyrophosphphorylase. In the course of this reaction there is the formation of a pyrophosphate (PPi) molecule.
c. The synthesis of the glycogen molecule is a bit unique in that the building of the chains requires already existing chains be present. The synthesis of the glycogen chain cannot be done de novo. Glycogen synthase is the enzyme that is responsible for the process by which the glycogen molecule

will grow by, but it requires the presence of the already existing glycogen molecule. This addition is accompoilished through the addition of the glycosidic bond between the anomeric carbon 1's hydroxyl and the carbon 4. This bond is the α-1,4 link that causes the elongation to continue. In humans the process of the growth of the chains is not a linear action, but rather a multiple branching. In the glycogen growth there is a branch that will occur approximately every 8th residue. This is a key element in the metabolic activity of humans as it will allow for the greatest possible number of nonreducing attachment points so the chain can grow in multitudes rather than at a single point. The enzyme that is responsible for the branching is the amylo-(α-1,4→α-1,6 transglycolase or the glucosyl α 4:6 transferase). The activity of this enzyme is the transfer of 5to 8 glucosyl residues from the non-reducing end to another residue at another site through the formation of the α -1,6 bond alluded to in the name of the enzyme. This process is repeated many times allowing the molecule to grow at very rapid rates.

d. Glycogenin is able to act as the initiation glycogen molecule even though it is a protein. This ability is due to the tyrosine residue hydroxyl present that can act as the receptor site for the glucosyl. This specific process itself has a responsible enzyme called glycogen initiator synthase, and this process takes place within the cytosol.

L. The degradation of the glycogen molecule is a more simple process but not a reversal of the synthesis. The process of this degradation is called glycogenolysis. In the degradation, the enzyme glycogen phosphorylase will break the α-1,4 bond by using pyridoxal phosphate as a cofactor to continue this process until there are only 4 glycosyl residues in the line. At this point there is a termination of the activity and the resultant structure is known as a limit dextrin and cannot be further acted upon by this enzyme.

a. The liberation of the branch itself is done through the activity of two enzymes that act in sequence. Glucosyl 4:4 transferase and amylo-α-(1,6)-glucosidase.
b. The glucosyl 4:4 transferase is responsible for the removal of 3 of the 4 glucosyl residues. These residues will be transferred to the non reducing end of another chain. The removal of these residues will leave a single glucosyl residue.
c. The remaining single residue will be acted upon by the enzyme amylo-α-(1,6)glucosidase.
d. Glycogen phosphorylase is now free to exert its effect on the remaining glycogen molecule. The product of this reaction is glucose 1 phosphate and this will be further acted upon as it moves further along the metabolic pathways.

M. Regulation of the delicate balance between the synthesis and the degradation is done through two specific mechanisms. One mechanism is the hormonal regulation between glucagon and epinepherine.

The other mechanism is in the allosteric activity of glycogen synthase and glycogen phosphorylase.

a. In the well fed state the body will signal the liver to increase synthesis while the opposite is seen in times of fasting. This signal is mediated by insulin. In the well fed state, the signal from insulin is to be anabolic and store things.
b. In the signaling mediated by glucagon or epinepherine, the activity is directed towards the activation of cAMP-dependent protein kinase. In this reaction there is a subsequent reaction that is either stimulated or inhibited as the end product of the reaction. The cAMP dependent protein kinase is a tetramer that will act upon the enzyme phosphoryl kinase via a phosphorylation reaction, making it an active enzymatic form from its inactive form. Glycogen phosphorylase can be acted upon by the phosphorylase kinase in its active form to be converted from glycogen phosphorylase b to the active form glycogen phosphorylase a. This is again due to a phosphorylation reaction. It is here that the degradation of glycogen will begin.

Citric Acid Cycle

I. This metabolic pathway is located in the mitochondria. More specifically it is found to lie within the mitochondrial matrix. This pathway is also considered by many as the final pathway for the oxidation of the amino acids, the carbohydrates and the fatty acids. This description is due to the main function of the Citric Acid Cycle AKA, TCA or Kreb's Cycle being the oxidation of Acetyl CoA to CO_2 and water. The TCA cycle has the ability to form glucose from the carbon skeletons of amino acids. This pathway is very intimately related to the glycolytic pathway as the end product of glycolysis becomes the beginning of the TCA. The reaction that is the linkage is as follows:

Fig 20

Pyruvate + CoA + NAD^+ → acetyl CoA + CO_2 + NADH

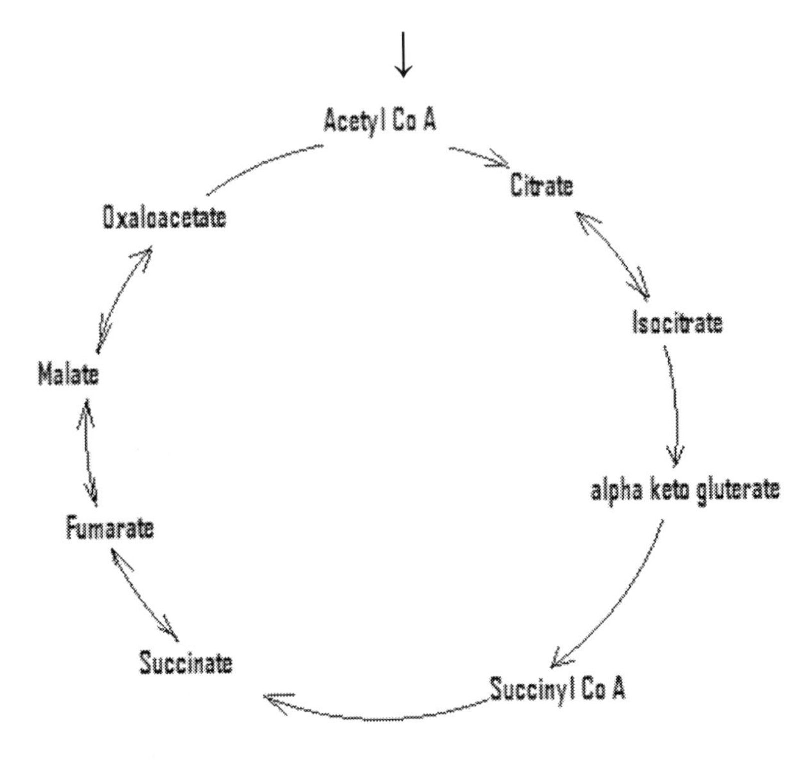

A. Oxaloacetate is a 4 carbon unit that is used with each turn of the cycle and it must first be condensed with acetate which is a 2 carbon unit for use in the cycle.

B. Pyruvate dehydrogenase is the enzyme that is responsible for the conversion of pyruvate to Acetyl CoA. The reactive portion of the Acetyl CoA is the acetyl group. The reaction will utilize H_2O and produce citrate and CoA. This enzyme is considered as a complex itself and it is not a part of the TCA proper. The reactions of this complex are as follows:

Pyruvate + CoA + NAD → acetyl CoA + CO_2 + NADH

This reaction is however requires several cofactors in order to drive it in the direction of forming acetyl CoA. Thiamine pyrophosphate is one of these cofactors. The requirement for this catalytic cofactor is the subsequent decarboxylation.

i. Pyruvate decarboxylase is responsible for the formation of the hydroxyethyl in the decarboxylation of pyruvate.

ii. Dihydrolipoyl transacetylase is responsible for the oxidization of the hydroxyethyl to yield lipoic acid. This is done via a transfer to a disulfide bond with the enzyme.

iii. Dihydrolipoyl dehydrogenase is responsible for the oxidizing of the sulfhydryl form of the lipoic acid. The yield of this reaction is an additional oxidized lipoic acid.

A. Oxaloacetate and Acetyl CoA will produce citrate with the enzyme citrate synthase. This reaction is an aldol condensation and it runs favorable in the direction of synthesis and not degradation.

i. In this reaction there is an intermediate formed, Citryl CoA.

ii. It is this intermediate that is responsible for the drive of the reaction in favor of the formation of Citrate.

D. Citrate, the six carbon tricarboxylic acid, will be acted upon by the enzyme aconitase in an isomerization reaction to yield isocitrate. The reaction is actually a two step process that first involves a dehydration and then a hydration. The enzyme responsible for this is called aconitase with cis-aconitase being the intermediate in the reaction.

i. This formation of citrate is a very important regulatory and control point of the cycle.

ii. ATP is an allosteric inhibitor of the enzyme citrate synthase.

iii. In this set of reactions, the exchange of the H with the OH is carried out through the dehydration and subsequent hydration reactions.

E. Isocitrate will be acted upon by isocitrate dehydrogenase in the oxidative decarboxylation of isocitrate to yield NADH and CO_2. This reaction is stimulated by things that signal low energy stores, increased levels of ADP and decreased levels of NADH. This will eventually lead to the production of α – ketogluterate.

i. This enzyme is also allosterically activated by the presence of large amounts of ADP

ii. Binding of NAD, Mg^{+}, and ADP are found to be mutually cooperative within the activity of this enzyme.

F. The conversion of α – ketogluterate to succinyl CoA is accomplished through the enzyme α – ketogluterate dehydrogenase complex.

i. This reaction will release the 2nd CO_2 and generate the 2nd NADH.

ii. There are a number of cofactors required to drive the reaction forward. Thiamine pyrophosphate, lipoic acid, FAD, NAD+ and CoA are required for this reaction.

iii. The activity of the α keto-gluterate dehydrogenase enzyme is known to be inhibited by the concentration of succinyl CoA and NADH.

H. Succinyl CoA will be cleaved by succinate thiokinase to succinate. This reaction releases GTP, a compound that is analogous to ATP. The energy yield is the same as for ATP.
Succinyl CoA + Pi + GDP ↔ Succinate + GTP + CoA

i. This is a substrate level phosphorylation.

ii. Succinyl CoA can be formed from fatty acids containing odd numbered carbons. Succinyl CoA can also be formed from propinyl CoA.

iii. Heme synthesis is dependent upon Succinyl CoA.

iv. Succinyl CoA synthase is the enzyme that is the catalyst for this reaction.

H. Fumarate is synthesized from the oxidation of fumarate using the enzyme succinate dehydrogenase. In this reaction, there is additional synthesis of $FADH_2$. The reducing power of succinate is not powerful enough to allow NAD as the reducing agent.

I. Fumarate is hydrated by fumarase to malate

J. Malate will then be oxidized to Oxaloacetate using malate dehydrogenase as the catalyst. This reaction will generate the final NADH of the TCA. This brings the reaction pathway full cycle.

K. The reaction accounting looks as follows

Acetyl CoA + 3 NAD + FAD + GDP + Pi + 2 H_2O → 2 CO_2 + 3 NADH + $FADH_2$ + GTP + 3 H^+ + CoA

L. ATP accounting is as follows:
Electron transport chain NADH → 3 adenosine triphosphate
Citric Acid Cycle $FADH_2$ → 2 adenosine triphosphate

M. Loss of four pair of hydrogen atoms, NAD^+ becomes reduced x 3, one FAD molecule is reduced, one high energy P is generated and two molecules of H_2O are hydrolyzed in the turning of this cycle one time.

N. Regulation of the Citric acid Cycle is accompoilished through the input of the 2 carbon bases as citrate from oxaloacetate. In this regulatory point ATP is important in the binding of the Acetyl CoA through saturation of the enzyme binding.

i. Isocitrate dehydrogenase is another regulatory point for the cycle. In situations where there is abundant ATP, the signal is to inhibit the drive of the reaction in synthesis.

ii. The final control point is found in alpha keto gluterate dehydrogenase as it is inhibited through increased levels of succinyl CoA.

Gluconeogenesis

I. Gluconeogenesis is the process of synthesizing carbohydrates from noncarbohydrate substrates. It is a process to manufacture glucose. 90% of gluconeogenesis is done within the liver. In addition to the liver, the kidneys are able to manufacture the remaining 10% of the glucose required. This is unusual situation in the absence of prolonged starvation. In this situation, glucose will be derived from lactate, glycerol, pyruvate and alpha ketoacids.

A. The precursor molecules are amino acids, glycerol, pyruvate and lactate.

i. Glycerol is obtained from the adipose tissue with hydrolysis and conversion to glycerol phosphate that will be oxidized to dihydroxyacetone.

ii. Lactate will be released to the body from cells that are devoid of mitochondria.

iii. Amino acids like pyruvate, Oxaloacetate and α – ketogluterate come from the glycogenic amino acids.

B. This biochemical pathway is located within the liver. The liver is responsible for 80 – 95% of the glucose synthesized in this pathway.

C. There are situations where other organs are capable of making glucose using this pathway.

i. Starvation is one of those extreme conditions. In this situation the kidney will produce a significant amount of glucose.

ii. The only other organ that is capable of this synthesis pathway is the small intestine. The epithelial cells of the small intestine are able to contribute a very small amount of glucose via gluconeogenesis.

D. The body utilizes gluconeogenesis in times of inadequate or low carbohydrate intake. In this scenario, the glycogen in the liver is insufficient to maintain adequate levels of blood sugar.

i. It is after 24 hours that the body will convert to this pathway.

ii. With prolonged bouts of strenuous exercise, the gluconeogenic pathway turns lactate and glycerol into a valuable energy sources. This is after the circulating levels of catecholamines have metabolized the carbohydrate and lipid stores.

iii. Lastly, the gluconeogenic pathway is a point of entry for the proteins to provide energy via the liberation of energy from the metabolism of the protein backbones for the amino acids to enter the metabolic pathways as urea.

E. Lactate is converted to pyruvate in normal metabolism via the Cori cycle in the Liver.

i. Glycolysis produces lactate from multiple reactions of the pathway on the initial glucose molecule. The final part of the pathway produces pyruvate form lactate using lactate dehydrogenase. This enzyme requires NAD+ as a cofactor.

ii. Lactate then becomes converted to the simple sugar glucose by gluconeogenesis and released from the liver to the blood stream.

iii. Tissues that are gluconeogenic have the equilibrium of the reaction running to formation of pyruvate. There are specific factors that will also help in the direction of the reaction and those are the isozyme of LDH and the ratio of NAD+ to NADH.

F. Pyruvate in muscle can be converted to Alanine through transamination that will release the Alanine into the blood to be used by the liver in gluconeogenesis.

G. Liberation of the fat cells will yield the glycerol backbones through the lipolysis of the triacylglycerol. In this reaction, the glycerol is modified in the liver through a phosphorylation reaction into 3 phosphoglycerate for use in the gluconeogenic pathway.

II. Reactions of Gluconeogenesis

A. As it is mainly a reversal of the glycolytic pathway, many of the substrates are alike. The steps of the reaction are also like with those non-reversible steps requiring additional reaction activity.

i. The phosphorylation of glucose by glucokinase

ii. The activity of phosphofructokinase as it converts fructose 6 phosphate (F6P) to fructose 1,6 bisphosphate (F1,6 BP)

iii. The activity of pyruvate kinase as it converts Phosphoenolpyruvate to pyruvate.

B. Pyruvate carbon dioxide and adenosine triphosphate are converted to an intermediate of the TCA cycle (Krebs's Cycle) by the action of pyruvate carboxylate. For this reaction to run there is a need for biotin to act as prosthetic group and for magnesium and manganese to act as cofactors. This reaction is activated by Acetyl CoA. This reaction is run within the mitochondrial matrix.

C. Oxaloacetate will be converted into Phosphoenolpyruvate via the activity of Phosphoenolpyruvate carboxykinase (PEPCK). This reaction is seen in the cytosol and requires manganese for activation.

D. Oxaloacetate can be converted into malate via the enzyme malate dehydrogenase in the mitochondria. There is additional activity of this reaction as a shuttle for reducing equivalents in and out of the cytosol.

  i. This reaction is imperative since Oxaloacetate cannot pass through the membrane of the mitochondria. Since the diffusional coefficient is not favorable it requires transport across the membrane as malate.

  ii. Once in the cytosol, the malate dehydrogenase will reverse the reaction and make Oxaloacetate for conversion into Phosphoenolpyruvate.

E. Phosphoenolpyruvate is converted to fructose 1,6 bisphosphate by the same six reactions that are seen in glycolysis and utilize the same enzymes as glycolysis.

F. Fructose 1,6 bisphosphate is converted to fructose 6 phosphate via the enzyme fructose 1,6 bisphosphatase

(the major regulatory enzyme of gluconeogenesis). This reaction is stimulated by citrate and inhibited by adenosine monophosphate and fructose 2,6 bisphosphate.

G. fructose 6 phosphate is converted to glucose 6 phosphate through the same enzymatic action as the reverse reaction in glycolysis, phosphoglucose isomerase.

H. Glucose 6 phosphate is converted to glucose and phosphate through the activity of glucose 6 phosphatase. This is not the same enzyme as is seen in the reverse reaction of glycolysis.

Lipids

I. Lipids are hydrophobic in nature and as such they are insoluble or very poorly soluble in water. Conversely, these are extremely soluble in nonpolar solvents like benzene, ether and chloroform.

II. Function of lipids as biologic lipids are as a storage form of metabolic energy and a transport form of metabolic energy. Biologic lipids are also important in the surface of the membrane of a cell. Biologic lipids are also structural components of cell membranes. Human dietary intake of lipids is in the range of 60 -150g/day, with close to 90% of those being ingested as triacylglycerol. The remainder is derived from the cholesterol, cholesteryl esters, phospholipids and the unsterified fatty acids.

III. Fatty Acids are long chain hydrocarbons that are insoluble in water. These molecules have a carboxyl group at one end and these molecules can be saturates or unsaturated.

A. Saturated fatty acid indicates that the chain has no double bonds.

B. The general formula for a saturated fatty acid is [CH_3—$(CH_2)_n$—COOH] so that n indicates the number of the methylene groups found between the methyl and the carboxyl carbons.

Fig 21

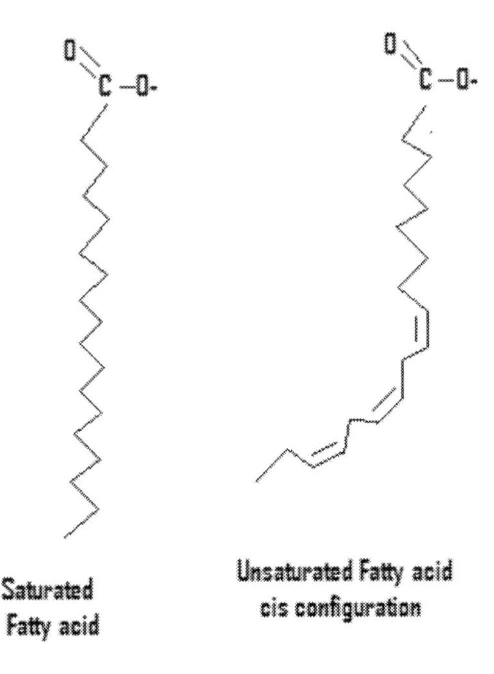

C. Unsaturated fatty acids have one or more double bonds. This system is identified by the use of the Delta system (Δ).
   i. In this system, the terminal carbon is carbon 1, with the double bond being given the number of the carbon atom on the carboxyl side of the double bond
   ii. The double bond in a naturally occurring fatty acid is always in the cis configuration.

D. Sources of nonessential fatty acids can be synthesized from the products of glucose oxidation.

E. Sources of essential fatty acids are required from the dietary intake. These are the linoleic ($18{:}2{:}\Delta^{9,12}$) and linolenic ($18{:}3{:}\Delta^{9,12,15}$).

i. There are no enzymes in human systems that are able to induce a double bond beyond carbon 9 of a fatty acid chain.
ii. All the double bonds that are produced are set 3 carbon atoms apart
iii. These properties along with the manner that a fatty acid is elongated by 2 carbon atoms, makes it impossible for the body to de novo synthesize polyunsaturated fatty acids.

F. Physical Properties of fatty acids are variable in the manner that they exhibit both polar (CH_3) and nonpolar (-COOH) terminal ends. This structure allows the polar heads to align with water and the nonpolar ends to align with the nonpolar (hydrophobic) side.
i. Melting points are determined by the chain length and by the number of double bonds. This translates into the melting point being higher with increasing length and the melting point dropping as the number of double bonds increases.
ii. The increased number of double bonds also will increase the fluidity of the membrane.

G. Triacylglycerides (Triacylglycerol) are trimesters of glycerol and three fatty acids. The functions of the triacylglycerols are as metabolic fuels and transport energy substances.

Fig 22

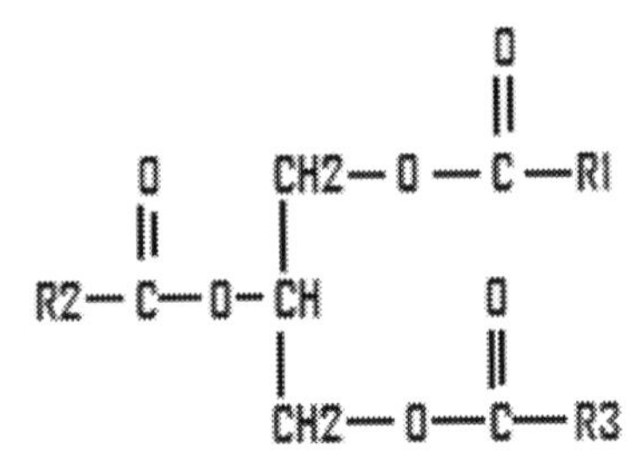

Triacylglycerol

i. Fatty acids are converted to triacylglycerols so that the fatty acids can be transported between tissues as they are to be utilized as metabolic energy substrates.
ii. The advantage of the triacylglycerol as a metabolic fuel is that the are a very concentrated form of energy as they are completely metabolized and combusted into CO_2 and water, 9 kcal/g versus the 4 kcal/g for carbohydrates. Additional benefits are that the triacylglycerides are not osmotically detrimental to the cell.
iii. Fatty acids in the blood as is seen in fasting, is due to the hydrolysis of fatty stores via lipolysis. Hydrolysis of the triacylglycerides will yield free glycerol and free fatty acids (nonesterified fatty acids). These levels of fatty acids will increase even with normal fast of sleep overnight.
iv. Almost every tissue will use fatty acids in some manner. The notable exception here is the brain.

The liver will use the glycerol that cannot be used by adipocytes will be transferred to the liver for gluconeogenesis.

v. Metabolism of the dietary lipids is a long and complex process. This process is initiated in the mouth via the action of lingual lipase. As previously mentioned, this emulsification is minute as compared to the remainder of the system. There is some digestion of lipids by the enzyme gastric lipase as it acts on the triacylglycerols associated in milk or other short to medium length chain fatty acids. In order for the digestion to begin, there is an additional protein, colipase that is secreted from the pancreas that is required for the stabilizing of the lipase in the point of attack, the lipid aqueous interface. The necessity for the emulsification is from the formation of the micelles that lipids produce. Without the emulsification, there is no ability for the enzymes to elicit their effects on the lipids. The duodenum is the structure that does the majority of the emulsification. The emulsification process is designed to increase the surface area as it is only possible to emulsify the portion of the micelle that is at the surface - water interface. This emulsification is done in two parts with the mechanism of action being the detergent like properties of the bile salts and the mixing of the components through peristaltic waves.

vi. The hydrolytic enzymes that have been mentioned previously are produced and secreted by the pancreas. The pancreatic manufacturing of these enzymes are regulated by hormonal signals. Cholecystokinin (CCK) or pancreozymin will be secreted into the jejunum and the duodenum as a result of the lipid and partially degraded proteins within the digestive tract. The target of these two hormones will be the gallbladder that will contract in response to the signal, release bile. Additional signals will be sent to the pancreas to release more digestive enzymes. These hormones will inhibit the peristaltic waves and slow down the motility of the lipids, thus increasing the time that the lipids are in contact with the digestive enzymes. Additional hormone participation will be from the hormone secretin. This hormone is released in response to the changes in pH within the intestine. The biochemical response of the hormone is to yield the pancreas to release a watery bicarbonate rich solution to buffer the acidic environment of the contents.

vii. The production of cholesterol and free fatty acids is due to the action of pancreatic cholesteryl ester hydrolase also known as cholesterol esterase. This enzyme will hydrolyze the cholesteryl esters. Digestion of the lipids is continued with the action of phospholipase A2. This enzyme is found in the pancreatic secretions. The action of this enzyme will be to remove the fatty acid from carbon 2 of the phospholipid. The removal of this fatty acid from carbon 2 will produce a lysophospholipid. Once this is completed, the remaining carbon's fatty acid will be hydrolyzed by lysophospholipase. The product of this reaction will be the generation of a glycerylphosphorylcholine. Clearance of the remaining compound will be through the urinary system as the glycerylphosphorylcholine.

viii. Dietary intake of lipids is a requirement for metabolic function. The compounds that are useful for metabolic intermediates are free fatty acids, free cholesterol, and 2 – monoacylglycerol. The ability of the body to resorb lipids is through the brush border of the intestinal membrane. Combination of the bile salts and the free fatty acids, free cholesterol, and 2 – monoacylglycerol become the components of the mixed micelles that move to the brush border for absorption. The mixed micelles will penetrate the unstirred water layer for absorption into the mucosal cells.

viv. The mucosal cells of the intestinal tract will be required to resynthesize the triacylglycerols and the cholesteryl esters. This is begun with the enzyme fatty acyl CoA synthetases, also known as thiokinase.

Fig 23

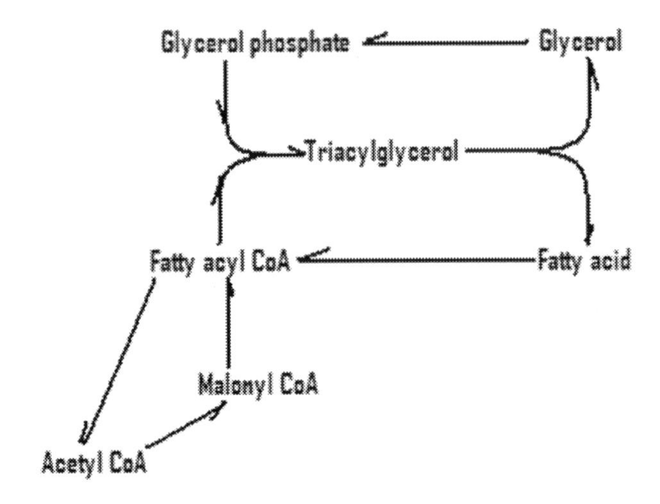

Triacylglycerol synthesis and degradation

H. Ketone Bodies are precursors of acetate, acetone, acetoacetate and B – hydroxybuterate.

i. These are synthesized within the mitochondria of the liver, to be used as metabolic fuels by every other tissue, except the liver.

ii. The synthesis process is restricted to periods of increased fatty acid oxidation coupled with decreased carbohydrate availability.

iii. In starvation, the production of ketone bodies will dramatically increase. This increase will be slow to activate, the maximal rate of synthesis will be seen at 20-30 days.

IV. Phospholipids are a major component of the cell membranes. This classification of lipids has a phosphorous molecule, a glycerol backbone or sphingosine.

A. Phosphoglycerides are composed of the following:
   i. Phosphatidylcholine (lecithin)
   ii. Phosphatidylethanolamine
   iii. Phosphatidylserine
   iv. Phosphatidylinositol
   v. Cardiolipin

B. These lipids are amphipathic – containing both hydrophobic and hydrophilic in nature.

C. These lipids are also amphoteric having both negatively and positively charged groups.

V. Sphingolipids are lipids seen in great concentrations in the CNS, especially the white matter. This lipid is also seen in other tissue as well. Classifications of sphingolipids are sphingomyelin and ceramide.

A. Sphingomyelin is a major component of neural tissue and membranes. This is the only sphingolipid that does not have a sugar moiety. The lipid is derived from ceramide (fatty acid and a sphingosine) and cytidine diphosphate (CDP-Choline).

B. Ceramides are a linkage of the fatty acids to the NH_2 group of the sphingosine through an amide linkage. This is a core structure of the naturally occurring sphingolipids.

C. Glycosphingolipids are a sphingolipid with an attached carbohydrate. There are subclassifications of these lipids that have been identified.
   i. Cerebrosides are seen in the myelin sheath and in other neural tissues. This classification of glycosphingolipid is a ceramide monohexoside.

Examples of this classification are the glucocerebroside and galactocerebroside.

ii. Globosides have two or more sugar molecules attached to the cerebroside. These sugars are galactose, glucose, N-acetylgalactosamine. These glycosphingolipids are seen in the spleen, the liver, and RBC's.

iii. Sulfatides are B – Sulfogalactocerebroside is a primary brain sulfolipid. It constitutes 15% of the white matter lipid content.

iv. Gangliosides contain one or more neuraminic acid residues. These are seen in higher concentrations in the ganglion cells and lower concentrations in other cell membranes. The letter "G" is the designate for a ganglioside. The subscripts M,D, T or Q are the designate for mono-, di -, tri -, or quarto – sailic acid components.

VI. Eicosanoids are the byproduct of the metabolism of the 20 carbon polyunsaturated fatty acid arachidonic acid.

A. Prostaglandins (PG) are analogs of prostanoic acid. These chemical compounds are found in many tissues and their function is not totally understood.

B. Functions of this compound are known to be smooth muscle relaxation and contraction, gastric acid secretion, Platelet aggregation, Inflammatory response, modulate sodium retention and water retention via the tubules of the kidneys.

C. Thromboxanes (TX) are analogs of prostanoic acid with a 6 member oxygen containing ring.

i. Subscript designates are indicative of the number of double bonds present.

ii. TX A_2 is responsible for the contraction of the smooth muscle within the tunica intima of the arteries. It is produced by platelets and it elicits platelet aggregation.

iii. Antagonist chemical compound to this action is the prostacyclin (PGI_2) produced from the endothelial cells.

D. Hydroperoxyeicosatetaenoic acids (HPETE's) do not contain a ring structure. This is a hydroxy fatty acid derivative. The function of these compounds is not totally understood but they can be converted and activated to compounds like leukotrienes.

E. Leukotrienes (LT) are derivatives from the HPETE's using lipoxygenase. The hallmark of this compound is a three conjugated double bonds.

i. These are all derived from arachnadonic acids.

ii. These are involved in the Chemotaxis, inflammatory response, and allergic reactions.

VII. Steroids are lipids containing four fused carbon rings. The formation of these fused carbon rings will produce cyclopentanoperhydrophenanthrene.

A. Sterols are classified according to the hydroxyl group at Carbon 3, aliphatic chain of at least 8 carbons at C 17.

B. Cholesterol is the major sterol of the human body.

i. Cholesterol is a structural component if cell membranes and plasma lipoproteins.

ii. This is a precursor of bile salts and steroid hormones.

iii. Steroid formation is done in the adrenal cortex, testies, ovary, and the placenta. The adrenal cortex produces 2 types of hormones, cortisol in the zona fasciculate and aldosterone that is produced in the zona glomerulosa. Cortisol inhibits the inflammatory response through a 11-keto derivative, cortisone. Aldosterone regulates the sodium absorption in the renal tubules. Aldosterone is a mineralcorticoid.

iv. Gonadal steroids produce testosterone through the Leydig cells of the testis. Ovaries produce estradiol in the thecal cells of the graafarian follicle and progesterone being formed in the corpus leuteum and placenta.

C. Bile Acids are cholic, chenodexoycholic, deoxycholic, and lithocholic acids. These compounds are C-24 steroids that are metabolized from cholesterol.

i. The cholic and chenodexoycholic are formed in the liver and are termed the primary bile acids.

ii. Deoxycholic acid and lithocholic acids are termed as secondary bile salts. Secondary bile are produced from the primary bile acids using intestinal bacterial enzymes.

iii. Bile acids become conjugated to glycine or taurine within the liver producing glyco taurocholate.

iv. The function of the bile salts is to convert the cholesterol to bile so that it will not accumulate within the tissue.

v. Bile salts are excreted in the feces.

VIII. Lipoproteins are complexes that contain proteins and lipids. These compounds are held together by noncovalent bonds. These are classified by either density or electrophoretic mobility.

A. Density exhibits high density (HDL), low density (LDL), intermediate density (IDL), very low density (VLDL), and chylomicrons.

B. These are spherical structures and they contain a protein and amphipathic lipid forming the outer shell. Internally, there are nonpolar lipids (TAG's and esterified cholesterol).

C. When treated with organic solvents, a lipoproteins will leave apoproteins.

VIV. Use of dietary lipids will be initiated as soon as the lipids enter the mouth. In the oral cavity, the enzyme lingual lipase will act upon the lipid once it enters the cavity. In comparison to the remaining digestive system, there is only a very minimal amount of digestion that is seen in the mouth. This digestion will continue as the lipids enter the stomach through the enzyme acid stable lipase. It is unlikely that there will be a significant amount of digestion of these dietary lipids in the stomach as the bulk of lipid digestion is seen in the small intestine. The digestion of these lipids will initially be slowed as the lipids are not yet emulsified. Emulsification of dietary lipids is done only at the lipid- water interface. Therefore, the emulsification of the lipid is a very important process.

X. Triacylglycerols are important storage units of the body. If there is excessive carbohydrate or protein intake through the diet, these compounds will be converted into triacylglycerols for storage. The synthesis of fatty acids is done primarily in the liver. Additional synthesis will be done in the mammary glands, adipose tissue and kidney's. Synthesis of these fatty acids will require the carbons that are to be donated for incorporation to be obtained from the Acetyl CoA. This is an energy intensive process that requires large amounts of ATP for hydrolysis. For the reaction pathway to complete its function, there is also requirement of large amounts of NADPH as the electron carrier.

A. The formation of these fatty acids is a complex reaction that has the requirement of movement of acetate from the mitochondrial side of the mitochondrial membrane to the cytosolic side. The acetate that is transferred is derived from the oxidation of pyruvate or the direct degradation of fatty acids, ketone bodies, or amino acids. The resynthesize of the acetyl CoA is done on the cytosolic side of the membrane. The reason that the direct transfer cannot be done is that the coenzyme A of the acetyl CoA cannot directly transfer across the membrane. This necessitates the transfer of the acetate is made with the acetate being in the form of citrate.

B. The storage of the monoacylglycerides, diacylglycerides and triacylglycerides will be into the adipose cell. The fatty acid on carbon 1 is generally saturated and the fatty acid on carbon 2 is generally unsaturated leaving the remaining fatty acid on carbon 3 to assume either configuration. Since the fatty acids are esterified through the carboxyl group, there is a net loss of negative charge. These TAG's are only slightly soluble in water and therefore cannot form stable micelles. Rather they will form oily micelles. These oily micelles are anhydrous, while providing the major energy source of the body. In order for the fatty acid to be converted to triacylglycerol storage, it must become activated. Fatty acyl CoA synthase or thiokinase is the enzyme that makes this activation happen. The defining aspect of activation is the attachment to coenzyme A.

i. the triacylglycerol is stored within the adipocyte in its anhydrous form.
ii. Storage is dependent on the combination with the cholesterol and cholesteryl ester, phospholipids and apo-B100 proteins. The outcome of the combination of these particles will produce a very low density lipoprotein or VLD for release into the blood stream for transport to the peripheral tissue. Travel in the blood stream only can occur if bound to albumin.

C. The release of the fatty acids for energy through hydrolytic cleavage will yield fatty acids and glycerol. For this release of energy there must be activation by hormone sensitive lipase.

Metabolism of Fatty Acids

I. Fatty acids are important resources in the body for a couple of very important reasons. They represent the easy energy source for the body as well as represent the main storage unit for excessive dietary sources.

A. Fatty acids are able to be synthesized from glucose through pyruvate. This synthesis is driven by the amounts of fat within the diet. Additional driving mechanism is the requirements for the conversion of glucose into a storage form

i. Synthesis is seen in the liver mainly.

ii. The enzymes that are necessary for the biosynthesis of these fatty acids are found within the cytosol. The enzymes required for the biosynthesis of fatty acids are not the same for the degradation of fatty acids in mitochondria.

B. Metabolically, fatty acids are the initial energy source for cardiac tissue.

C. Synthesis of fatty acids is accomplished through the polymerization of 2 carbon units that are derived from acetyl CoA (acetyl coenzyme a). The end product of this pathway will be the 16 carbon palmitic acid, a saturated fatty acid. Cofactors for this pathway are ATP and NADPH.

i. Acetyl CoA is the end product of the oxidation of glucose through glycolysis and pyruvate dehydrogenase.
ii. NADPH is derived from the pentose phosphate pathway
iii. Mitochondrial Acetyl Co A is not able to cross the inner mitochondrial membrane. Because of this inability of the Acetyl CoA to cross the membrane a mechanism must come into play with respect to a transport mechanism. This mechanism is the Acetyl CoA shuttle. As a byproduct of this reaction, NADPH is also produced.
iv. Reactions of the production of the NADPH of the Acetyl CoA shuttle follow these reactions:

Acetyl CoA reacts with oxaloacetate forming citrate by the enzyme Citrate Synthase. This reaction is run in the mitochondria. The citrate is then transported into the cytosol via tricarboxylic acid transport. The citrate will react with the CoA yielding acetyl CoA and oxaloacetate. This reaction is catalyzed by citrate lyase and will require the additional energy input of ATP. Oxaloacetate will then be converted to malate by malate dehydrogenase. The malate will then be decarboxylated to yield pyruvate by the malic enzyme. It is in this step that the formation of NADPH is produced from NADP+. Lastly, pyruvate will be transported into the mitochondria via active transport mechanisms. Concurrently, oxaloacetate will be regenerated from pyruvate using ATP and pyruvate carboxylate.

II. Reactions of the synthesis of fatty acids.

A. Acetyl CoA + HCO_3^- + ATP yield malonyl CoA +inorganic phosphate (Pi) catalyzed by Acetyl CoA carboxylase.

i. This is the regulatory site for fatty acid synthesis

ii. Feed forward by Citrate & Insulin.

iii. Negative feedback by palmitoyl CoA & Glucagon.

B. Acetyl CoA +acyl carrier protein (ACP) yields acetyl-ACP + CoA catalyzed by fatty acid synthase enzyme complex.

C. Malonyl CoA + ACP yields malonyl-ACP + CoA by the enzyme malonyl transacylase.

D. Malonyl –ACP + acetyl-ACP yields acetocaetyl-ACP +ACP + CO_2 by the enzyme 3-ketoacyl synthase

E. Acetoacetyl- ACP yields 3-hydroxybutyryl – ACP through the enzyme 3-ketoacyl reductase.

F. 3-Hydroxybutyryl – ACP yields crotonyl –ACP by the enzyme 3-hydroxybutyryl-ACP dehydratase

G. Crotonyl-ACP yields butyryl-ACP by the enzyme enoyl-ACP reductase using NADPH as cofactor.

H. This reaction will turn seven times to produce the palmitic acid, a 16 carbon fatty acid. The enzyme palmitoyl decyclase will release the palmitic acid from the ACP.

I. Additional carbon units are added by the elongation either in the endoplasmic reticulum or through the mitochondrial elongation system.

i. The mitochondrial elongation system will require the use of acetyl CoA. This is in contrast to the malonyl CoA described below. Fatty acids elongated by this mechanism are earmarked for structural lipid components.

ii. The endoplasmic reticulum is the most used elongation mechanism. It utilizes malonyl CoA units that are added to the traditional palmitate as in the formation of fatty acids described previously. In this elongation mechanism, CoASH rather than the ACP is the carrier. Steric acid, an 18 carbon fatty acid is a common end product of this mechanism of elongation.

III. Oxidation of Fatty Acids as an energy source is very important as it typically yields high energy compounds of $FADH_2$ and NADH during the reaction and yields acetyl CoA.

A. B –Oxidation of fatty acids is the main mechanism for the breakdown of fatty acids. This process is seen in the matrix of the mitochondria.

B. The beta oxidation of fatty acids comes from the oxidation of the beta carbon into a beta keto acid. This is accomplished by the uptake of the free fatty acids being removed from the circulation and internalized to be activated into the acyl CoA derivatives.

i. Activation of the fatty acids by the thiokinase inner mitochondrial acyl CoA synthases on the medium length chains, 4-10 carbons and the short chain, acetates and propinates is why these chains are able to enter the mitochondria and the long chains are not able to enter.

C. Carnitine (beta hydroxy-γ triethylamonium butyrate) is the mechanism of entry for the long chain fatty acids. Carnitine is synthesized from Lysine. It is found in the inner mitochondrial membrane and it role is to receive the fatty acyl group as it is being moved into the cell. It is this mechanism that is the basis for the carnitine shuttle allowing the entry of the long chain fatty acids for oxidation. The shuttle mechanism is found as components:

i. carnitine palmitoyl transferase I (CPT I) is the enzyme that catalyzes the transfer of the acyl group to the carnitine molecule. This transfer allows the translocation to occur.

ii. Translocation is completed and then the enzyme carnitine palmitoyl transferase II (CPT II) accepts the acyl group of the CoA from the matrix.

D. The pathway of this beta oxidation is a repeated four step processing of the fatty acid, cleaving 2 carbon units at a time until there are only 2 carbon units remaining.

i. Acyl CoA => enoyl CoA is the reaction catalyzed by the enzyme acyl CoA dehydrogenase that requires FAD as the prosthetic group.

ii. Enoyl => 3 hydroxyacyl CoA catalyzed by the enoyl CoA hydratase enzyme. This is as the name implies a hydration reaction.

iii. Hydroxyacyl CoA => 3 ketoacyl CoA reaction is catalyzed by the enzyme 3 hydroxyacyl CoA dehydrogenase. This is an important step as NAD+ is produced during the reduction of NADH.

iv. 3 Ketoacyl CoA + CoASH => acetyl CoA + tetradecanoyl CoA that is catalyzed by B ketothiolase. This is a thiolytic cleavage reaction.

IV. Oxidation of Unsaturated Fatty acids constitute almost half of the fatty acids found in humans. In naturally occurring fatty acids, the double bonds are found in the cis configuration. This is problematic for the B oxidation pathway of humans as they are not able to metabolize this configuration. The tans configuration is the configuration that is acceptable in the B oxidation pathway, therefore isomerase shifting is a requirement for the reaction to proceed. The enzyme responsible for this is Δ^3- cis- Δ^2 –trans enoyl CoA isomerase. The function of this enzyme is to isomerize the bond to the preferred Δ^2-trans configuration. If the fatty acid is a polyunsaturated fatty acid then additional enzymatic support is required in the form of epimerases to convert from "D" form to the "L" configuration.

V. Alternative metabolic pathways for fatty acid oxidation.

A. ω Oxidation is the oxidation of the terminal methyl group. This reaction will yield a ω fatty acid.

B. α Oxidation is the oxidation of long chain fatty acids to 2 hydroxy fatty acids. The 2 hydroxy fatty acids are the main brain lipid fatty acids. This reaction is followed by an additional oxidation reaction that removes one additional carbon.

VI. Triacylglycerides (TAG)

A. Fatty acids are converted to triacylglycerides for the transport to tissues and for storage in adipose cells. This requires the acylation of the three hydroxyl groups of glycerol. For this to occur, the fatty acid must first be activated. This activation occurs in the cytosol and is catalyzed by the enzyme acyl CoA synthase; this is an ATP dependent enzyme. There are two acylation reactions that take place.

i. First acylation of the first hydroxyl uses the dihydroxyacetone phosphate (DHAP) that will be followed by a reduction reaction to yield lysophosphatidate. There is a variable pathway to give the same end result but uses the reverse order.

ii. the second acylation is of an unsaturated fatty acyl thioester will be added to the 2-hydroxyl of the lysophatidate.

iii. Third acylation is when the phosphate of C3 is cleaved by the phosphatases that then will add a fatty acid, either a saturated or unsaturated fatty acid to the C3 hydroxyl.

B. Storage of triacylglycerides in adipose tissues is esterified for the storage. This process is dependent on the carbohydrate metabolism in glycolysis. This is because the formation of DHAP and glycerol 3 phosphate. Adipocytes are devoid of the enzymes responsible for the phosphorylation of the glycerol. This is due to the lack of the enzyme glycerol kinase. The entry of the glucose molecule into the adipocytes is dependent on insulin as well. This makes insulin imperative for the storage to occur.

C Lipolysis of triacylglycerols is able to be achieved via three pathways. Enzymes responsible for this hydrolysis are hormone sensitive triacylglycerol lipase, that catalyzes the reaction to yield diacylglycerol + free fatty acid. The next enzyme is diacylglycerol lipase. This is the enzyme that catalyzes the reaction eliciting the monoacylglycerols + free fatty acids. The remaining enzyme is the monoacylglycerol lipase enzyme. This enzyme is responsible for the final product of glycerol and free fatty acids.

i. Hormone sensitive lipase is the rate limiting step in this reaction. Activation of this enzyme is accomplished by cAMP and norepinepherine. Inhibition of this enzyme is seen by insulin.

D. The major form of storage of triacylglycerides is in adipocytes, however, muscle, and liver will store small droplets as intercellular lipids for use in that specific tissue. These lipid droplets are hydrolyzed by the same mechanisms as adipocytes.

VII. Ketone bodies are the main form of energy delivery to the heart skeletal muscle and kidney. In starvation, after 20 days, ketone bodies are the main form of energy for the brain. Synthesis of ketone bodies come from 2 acetyl CoA molecules, with a third acetyl CoA reacting with Acetoacetyl

CoA yielding B hydroxy – B methylglutaryl CoA (HMG CoA). HMG CoA is acted on by the enzyme HMG CoA lyase producing acetoacetate and acetyl CoA.

A. Activation of Ketone bodies for use as energy source. This is done by forming acetoacetate. This is catalyzed by 3 keto acid CoA transferase or acetoacetyl CoA synthetase.

B. This reaction will yield 2 acetyl CoA molecules, high energy phosphate compounds.

C. Each mole of acetyl CoA yields 12 moles of ATP through the TCA, electron transport chain, and oxidative phosphorylation. Total yield is 23 moles from oxidation.

II. Oxygen transport proteins – Oxygen delivery is dictated by the diffusional coefficient of oxygen as it diffuses across the cell membrane. Distance is also a limiting factor for this diffusion. It is for this reason that it appears the evolution of oxygen binding proteins or oxygen carrier proteins have developed. In the human body, these proteins are myoglobin and hemoglobin.

A. Myoglobin is the oxygen binding protein seen in muscle. The function of this protein is to transport oxygen to the mitochondria.

B. Hemoglobin is the oxygen binding protein that is responsible for the transport of oxygen to the tissue from the lungs and transporting the carbon dioxide and protons along with other waste products to the lungs. Hemoglobin is found within the red blood cells and it is very abundant with more than 280 molecules of hemoglobin per red blood cell.

III. Mechanism of action for Myoglobin and Hemoglobin

A. Myoglobin works by at the low oxygen tensions in the tissues. This is the point of release of the oxygen from hemoglobin.

B. The oxygen binding curve (oxygen dissociation curve) for myoglobin is a rectangular hyperbola.

i. The binding if myoglobin is 90% saturated when the pressure is 20 mmHg (torr) which is the Po_2 of the muscle.

ii. Myoglobin is a single peptide chain that consists of 153 residues.

iii. The configuration of this is with the majority (75%) of the helix arranged as eight alpha helical segments. These segments are designated A-H and they have polarized properties. The polarized helices (hydrophilic) are found on the outside of the molecule while the nonpolar (hydrophobic) helices are found in the inside of the molecule.

iv. The prosthetic group associated with myoglobin is heme. It is the heme that creates the oxygen binding property. Heme is a molecule that is derived from ferrous iron (Fe^{2+}) with a protophorphyrin ring. This protopophyrin ring is called protporphyrin IX. This molecule is made from iron as it binds to the central protporphyrin IX ring. The iron is bound by four nitrogen molecules.

v. Oxygen binding site is unusual in its make up as it has a ferrous heme molecule that is able to form six ligand bonds. Histidine makes the bond at the 5th position located on the proximal side of the protporphyrin ring plane while the 6th position is an oxygen molecule and an iron and histidine molecule located on the distal side of the ring.

vi. There is a hydrophobic pocket on the myoglobin that is responsible for the protection of the ferrous form of the iron from being oxidized to the ferric form of the iron ($Fe^{+3)}$ Should this oxidization occur, then the oxygen is replaced by a water and the molecule becomes metmyoglobin. The reason that the molecule will not bind oxygen is the additional electron will create a steric hindrance that will prevent the oxygen binding conformation from taking place.

C. Hemoglobin functions at the higher Po_2 in the lung. It binds oxygen and then carries it to the tissue where the pressures are lower and it is there that the hemoglobin releases the oxygen.

i. The oxygen binding curve for hemoglobin is sigmoidal in shape.

ii. Hemoglobin is a tetramer of four subunits with a hydrophobic pocket. The prosthetic group of hemoglobin and myoglobin are identical.

iii. Genetic variations of hemoglobin:

1. Hemoglobin A1 is the major subunit seen in adults. It is a tetramer of 2 identical alpha subunits and 2 identical beta subunits ($\alpha_2 \beta_2$)
2. Hemoglobin A2 is a minor subunit that is a tetramer of 2 alpha subunits and 2 delta subunits ($\alpha_2 \delta_2$)
3. Fetal hemoglobin: (HbF) fetal subunits are very different from adult.
    a. The first to develop is the 2 zeta units and to epsilon units. (ζ2 ε2) The zeta is analogous to the alpha and the epsilon is analogous to the beta.

b. At the end of the 6th month the zeta is replaced by the alpha but the epsilon units are replaced by the gamma units (ά2 γ2).

c. Just after birth the gamma units are replaced by the beta units of adult configuration.

d. Fetal hemoglobin has a higher affinity for oxygen than adult hemoglobin does.

e. The gamma units do not bind the 2,3 BPG well and therefore shifts the oxygen binding curve to the left.

D. The binding of the oxygen is a cooperative binding. Hemoglobin with no oxygen bound is called deoxyhemoglobin. The mechanism of this nonbinding is the steric hindrance of the proximal histidine and the distal nitrogen. Once the oxygen binds the iron atom is pushed into a planar configuration and the binding becomes easier.

i. The binding results in the movement of the iron into the planar configuration and this allows the proximal histidine into the heme. This change in position of the heme allow the first oxygen to bind. The binding of the first oxygen makes it easier for the remaining oxygen molecules to bind.

ii. Forms of Oxygen - deoxyhemoglobin is considered to be the tight or the T form. The binding of the first oxygen creates a steric change on the arrangement that the hemoglobin changes into the relaxed or R form.

IV. The Bohr Effect for Hemoglobin states that there is a intimate relationship between the hemoglobin oxygen affinity, Pco_2 level and pH.

A. The affinity of oxygen binding is decreased by increasing the hydrogen concentration, or increased Pco_2. This is demonstrated by the dissociation curve shifting to the right. Protons will bind to a specific site on the hemoglobin. The carbon dioxide binds to the terminal amino group as a carbamate. The binding of the carbon dioxide is greater in the T state than the R. The binding of the carbamate shifts the proton concentration to increase. This increased proton concentration stimulates the Bohr effect.

B. 2,3 BPG is a protein that is found within the erythrocyte in the same concentrations as the protein hemoglobin. The sole function of the 2.3 BPG is to lower the oxygen affinity. Increased levels of 2,3 BPG will shift hemoglobin into the T form. This will assist the hemoglobin to release oxygen in a more efficient manner.

V. Clinical Implications of Hemoglobins

A. Diabetes is the glycosylation of HbA_1 that produces a variation HbA_{1C}. This subtype is elevated in the serum of patients with Diabetes (6 -12%).

B. Hemaglobinopathies – mutations of the hemoglobin such that pathologic conditions present.

i. HbS – Hemoglobin S – Sickle cell anemia – valine replaces glutamic acid in the sixth position of the Beta chains.

1. this is the most common hemoglobinopathy
2. deoxygenation produces polymerization of the hemoglobin and this produces the sickle shape. These individuals have increased resistance to malaria.

ii. HbC –Hemoglobin S - the glutamic acid at position 6 on the Beta chain is mutated to Lysine. This produces a crystallization within the cell and not a sickle configuration. This is a mutation that is expressed phenotypically in blacks.

iii. HbM - Hemoglobin M – the proximal or the distal histamine are mutated. These mutations produce a stabilized form of ferric iron that cannot bind to oxygen. This yields a methemoglobin. This is a rare mutation. Homozygous individual are afflicted with the lethal form of the mutation.

C. Thalassemias are a form of hemolytic anemia. The mechanism of this pathology are from the lack of alpha and or beta chain production.

i. Thalassemia minor is the most common type of thalassemia. It is a heterozygous mutation of the beta chain. These patients have normal life spans with mild anemia. The compensatory mechanism is in the overproduction of the HbF and HbA_2 subunits.

ii. Thalassemia major is a rare mutation to be expressed in adulthood as it is a lethal mutation. These individuals are homozygous for mutated beta globin chains. These individuals have very severe anemia.

iii. Alpha thalassemias are of 4 major subcategories:

1. if there is only mutation in the expression of a single alpha subunit the individuals are normal phenotypically and are only carriers of the trait.
2. If there is deficiency of two of the alpha subunits, the individual is diagnosed as having alpha

thalassemia trait. These individuals will exprcss only mild anemia.

3. If there are mutations in three of the alpha globin subunits, the individual is diagnosed with Hb H disease. These individuals will exhibit severe anemia.
4. If there is mutation in all four of the alpha globin subunits, this condition is incompatible with life. Diagnosis is termed Hydrops fetalis. Individuals are either still born or death immediate post partum.

VI. Collagen is the most abundant of all human proteins. It is the major extracellular structural protein. Collagen is a class of proteins that carries with it at least 12 subtypes.

A. Type I collagen is a triple helix

B. Amino acid composition demonstrates at least one 1/3 of the total side chains being Glycine, 10% Proline, 10% hydroxyproline and the remaining 1% hydroxylysine.

C. The structural unit of collagen is tropocollagen. These subunits of tropocollagen crosslink to form the larger fibers. The tropocollagen is made from 3 polypeptide chains called alpha chains.

i. Unique characteristic of this configuration is that every third amino acid in this sequence is Glycine. This sequencing (gly-pro-X) or (gly-X-hyp) makes up approximately 60 % of the amino acid chain.

ii. Another unique characteristic of this configuration is that the alpha chain forms a left handed helix. This characteristic is unique on several manners, the fact that glycine makes up every third residue, means that the chain can turn tightly and this leads to interchain hydrogen bonding with the glycine's NH in the peptide bond to the carbonyl (C=O) groups in adjacent peptides that stabilize the structure. This means that the overall structure is a right handed triple helix made from three left handed alpha helices.

iii. Synthesis begins in the Endoplasmic Reticulum as translation into mRNA in the RER (rough endoplasmic reticulum). The first set of alpha chains produced are in the form of pre-procollagen. The final processing of the chains is with the 150 residues on the N terminus and the 250 residues on the C terminus being cleaved from the molecule. These signal sequences are required for the proper manufacturing signaling but will be cleaved to produce proper conversion from procollagen to collagen. This is post-translational modification of the alpha helices that takes place in the endoplasmic reticulum and must be completed for the proper designation of the collagen.

iv. Further processing of the procollagen by three specific enzymes. These enzymes require cofactors of Fe^{2+}, ascorbic acid, oxygen, alpha ketogluterate.

1. lysyl hydroxylase that converts the lysines in the sequence X-lys-gly to 5-hydroxylysine.
2. Prolyl-4-Hydroxylase that converts prolines in the sequence X-pro-gly to 4-hydroxyproline.
3. Prolyl – 3-hydroxyase that converts proline in the sequence Hyp – Pro-gly to 3-hydroxyproline.

v. Glycosylation is also required for the manufacturing of a stable triple helix. In this reaction, glactose is removed from the Uridine diphosphate galactose and it is added to hydroxylysly residues using the enzyme galactosyl transferase. Glucose is cleaved in the same manner using the same path but the enzyme is glucosyl transferase.

vi. Cysetine residues will form interchain disulfide bonds that add tremendous specificity of alignment.

vii. Once these steps are completed, the molecule is transferred to the golgi body for post-translational modification (O-linked glycosylation) with the procollagen molecule packed into secretory vesicles.

viii. Upon the secretion of the procollagen molecule, the N & C terminus is cleaved (propeptides) by specific procollagen peptidases that allow the formation of the tropocollagen. The tropocollagen molecule is very unique in the manner it actually forms via self assembly. The formation of collagen continues as each tropocollagen molecule aggregates with 4 other topocollagen chains to form a collagen fibril. These fibers will aggregate and form a collagen fiber. The strength of this structure is enhanced by the crosslinks of the aldehydes of the lysine and the hydroxylysine with the enzyme lysyl hydroxylase.

VII. Muscle Proteins allow the movement of muscle tissue. This is called the sliding filament theory of muscle contraction.

A. Striated muscle is composed of muscle fibers. Each muscle fiber is a single multinucleate cell that develops from myoblasts. Muscle fibers are composed of many myofibrils that can measure 1-2 μm in diameter. These fibers are ensheathed by the sarcolemma which is a plasma membrane with electrical conduction properties of excitability. Within the sarcolemma is the sarcoplasm, the analogous cytoplasm of other cell types. Attached to the sacrolemma via the T tubule system is the sarcoplasmic reticulum.

i. The functional unit of the muscle cell is the sarcomere. It is this unit that creates the distinguishing pattern of dark and light bands.
ii. The A band is the band that appears dark in the microscope.
iii. Within the A band is the lighter and less dense H zone. It consists of thick filaments that will extend the length of the A band. These fibers do not change length or width during a muscle contraction.
iv. In the middle of the H zone is the dark M line. This M line consists of M protein. This M line joins the fibers that make up the A band.
v. The I band is the lighter region on the microscope. Included with this structure is the dense, long Z line that delineates the margins of the sarcomere. The Z line consists of proteins alpha actin and desmin.
vi. Thin filaments consist of actin, tropomyosin, troponin. These filaments extend from the Z line and across the I band and through the A band to the H zone. These do not change length or width with muscle contraction either.

vii. The overlap of the thick and thin bands is seen as they interact within the A band. The region of this overlap is the darkest seen in the microscope.

viii. The H zone is the portion of the A band where there is no overlap of the thick and thin filaments. It narrows with contraction.

B. Thick Filaments are composed mainly of myosin. Myosin type 2 is the only form of myosin found in muscles. This is a large molecule consisting of two identical heavy chains and two different pairs of light chains. The heavy chains are attached via the coiling of the alpha helical tails forming the coiled coil structure.

i. The myosin protein is able to be enzymatically separated from other subtypes. The NH_2 terminal globular end is able to bind to actin as is demonstrated by the A band that results form the forming of the cross bridges of the myosin to the actin

ii. The amino terminal globular head of myosin exhibits ATPase activity

C. Thin Filaments as mentioned above consist of actin, tropomyosin, and troponins. Actin is a polymer of G actin subunits, each with a single molecule of ATP bound to the G actin.

i. G actin will polymerize to form filaments of F actin (the helical structure of linked G actin molecules)

ii. The hydrolysis of ATP in the conversion of G actin to F actin is not required for this polymerization to occur. It only enhances the rate of the formation of the F actin. The ADP product of the reaction is bound to the F actin and stabilizes it.

iii. Actin is the most abundant protein in the cell and has tropomyosin molecules many isoforms.

D. Tropomyosin consists of alpha and beta subunits that twist around each other to form a coiled molecule. The tropomyosin molecules occupy a groove in the actin filament. There is one tropomyosin molecule for every seven actin monomers. The tropomyosin is important in regulating the interaction of the actin and myosin filaments

E. Troponin is a protein with three isoforms. The function of troponin is to work with tropomyosin in the regulation of the contraction, but by the regulation of calcium.

i. Troponin T binds to tropomyosin

ii. Troponin I binds to Actin

iii. Troponin C binds to calcium

VIII. Mechanism of Contraction

A. Sliding filament theory – each sarcomere of a myofibril shortens with contraction as the thin filaments slide over the thick filaments.

B. The cycle of contraction is as follows:

i. Myosin with the ADP + Pi diffuses and contacts the actin, losing the Pi with the initial binding and myosin binds tightly to the actin.

ii. This tight binding creates a conformational change in the myosin that results in the pulling of the actin across the thick filament while releasing the ADP.

iii. ATP present causes the release of the myosin from the actin

iv. Myosin hydrolyzes the ATP to ADP + Pi.

C. In striated muscle, the regulatory chemical is calcium. In the absence of calcium, tropomyosin will lie in the groove of the actin helix such that the myosin – ADP-Pi complex cannot bind. If calcium is available, it will bind to troponin C that will elicit a conformational change in the whole molecule and clearing the tropomyosin from the myosin binding site on the actin molecule allowing myosin to bind and the contraction to continue.

i. In resting muscle, the calcium is stored in the sarcoplasmic reticulum

ii. Electrical stimulation elicits depolarization and this is transferred from the T Tubules to the sarcoplasmic reticulum, releasing calcium.

iii. Enzymatic cleavage of the calcium is completed and the reversal of the conformational change is begun.

Glycosaminoglycans

I. Glycosaminoglycans are integral components of many aspects of the human body. These are negatively charged heteropolysaccharide chains bound with a protein complex The importance of this charge and structure is in the ability of the molecule to bind tremendous amounts of water. The function of the glycosaminoglycans are in structural support and assist in the maintaining of the homeostasis of electrolytes within the tissue. There are six major classifications of glycosaminoglycans that are described according to their monomeric composition, number of sulfates, location of those same sulfates and the type of glycosidic bonds.

a. The charge is derived from the acidic sugar that is then bound to the acetylated sugar. The general structure of these compounds will be a long linear molecule of repeating disaccharides that consist of the acidic –amino sugar orientation mentioned above. In this configuration, the components of the amino sugar will generally be see as D-glucoseamine or D-galactosamine. In most scenarios, the sugar would be positively charged but in this situation, the sugar is seen to have a acetylated and this will remove the positive charge. The acidic sugar is generally known to be either D-glucuronic acid or L-iduronic acid. The sugars also contain sulfate groups that are contributors to the overall negative charge. The sugars are also known to have negatively charged carboxyl groups at physiologic pH and it is this compilation of factors that will provide the overall negative charge of the glycosaminoglycan.

b. The large number of negatively charged sugars will allow the increased retention of water in the tissues and as was demonstrated in basic chemistry, like charges will repel each other, a property that the body will take advantage of

c. The bonding of the glycosaminoglycan to the protein will create a proteoglycan monomer that is generally described as a core protein with linear carbohydrates covalently bound to it. This is a structure that is imperative for the proper function of cartilage. It is often described as a "bottle brush configuration"

d. The most common structural components of the glycosaminoglycans are

    i. Heparan sulfate a disaccharide with acetylated glucosamines, generally found in basement membranes and cell surface receptors.

    ii. Heparin which is best known as an anticoagulant. This is a disaccharide of glucosamine and glucuronic acid or iduronic acid in which the residues are sulfamide linkages. Heparin is found within mast cells therefore it is active in immune responses.

    iii. Chondroitin 4 sulfate is responsible for the majority of the glycosaminoglycans in cartilage tendons and ligaments. This is a disaccharide of

N-acetylglactosamine and glucuronic acid with the sulfate at C4.

iv. Chondroitin 6 Sulfate is also found extensively in the tissues. It is responsible for the strength of the collegan molecule.

v. Dermatan Sulfate is seen within the skin and the vessel walls. The structure of this molecule is repeating units of N-acetylgalactosamine and L-iduronic acid.

II. Synthesisoftheglycosaminoglycansoccursintheendoplasmic reticulum with final modification in the golgi complex. The activity of the synthesis of the glycosaminoglycans within the connective tissue can be defined through staining techniques as the synthetic pathway is very active within the connective tissue.

a. In the synthetic pathways for N-acetylglucosamine and N-acetylglucosamine, glutamine is the donor of the nitrogen. In this pathway, fructose 6 phosphate is the key sugar as it is the acceptor of the amide nitrogen to yield glucosamine 6 phosphate and glutamate. In an isomerization reaction, the glucosamine that was acetylated in the previous step will produce N-acetylglucosamine 1 P. The product of is then able to be available for incorporation into the growing unit.

b. The terminal carbohydrate on the oligosaccharide or possibly glycosaminoglycans is generally N acetylneuraminic acid (NANA). The synthesis of the NANA's is developed from the donation of the carbons and the nitrogens from acetylated mannose sugar residues (N-acetylmannosamine or phosphoenolpyruvate) This is a unusual reaction in that the form of NANA described is in itself inactive and therefore must be converted to an active form through a reaction that involves cytosine trisphosphate. This reaction is itself unusual in that the cytosine triphosphate is not the active high energy phosphate compound. CTP will be converted to cytosine monophosphate in a reaction that is catalyzed by N-acetylneuraminate – CMP – pyrophosphorylase

III. Synthesis of the acidic sugars is best described with the formation of Glucuronic acid. It can be formed from the degradation of the glycosaminoglycans or through the uronic acid pathway as it is metabolized and then enter the hexose monophosphate pathway. The end product of this reaction are the intermediates of glycolysis , fructose 6 phosphate and glyceraldehyde 3 phosphate. . ///////

Hexose Monophosphate Pathway

I. The hexose monophosphate pathway does not produce or use adenosine triphosphate in the execution of the reactions. This pathway is slightly different from other pathways as the actual reaction directions are ill-defined. This is in stark contrast to the glycolysis, TCA or gluconeogenesis. This pathway is found in the cytosol. The function of this pathway is the production of NADPH, a chemical reductant. The other function of this pathway is the production of the ribose phosphate required for the nucleotides and finally, it functions as a metabolic use for the dietary 5 carbon sugars. There are 2 sets of reaction types, the oxidative reactions and the nonoxidative reactions.

A. In the oxidative reactions, for each mole of glucose 6 phosphate that is oxidized there will be one mole of ribulose 5 phosphate, 2 molecules of NADPH and one mole of CO_2.

i. Glucose 6 phosphate dehydrogenase is the enzyme responsible for the reaction that converts glucose 6 phosphate to 6 phosphogluconate in an oxidation reaction.

ii. This is the main regulatory step of the pathway.

iii. The oxidative decarboxylation of the 6 phosphogluconate is accomplished via the enzyme 6 phosphogluconate dehydrogenase will yield a pentose sugar – phosphate, carbon dioxide and the 2nd NADPH.

B. The nonoxidative reactions will catalyze the interconversions of the 3, 4, 5, & 7 carbon sugars. This is important in how the ribulose 5 phosphate will be converted to the ribose 5 phosphate or even to the intermediates of glycolysis, like fructose 6 phosphate or glyceraldehyde 3 phosphate. The coenzyme that is required for this reaction to run is thiamine pyrophosphate.

II. NADPH is actually considered a high energy molecule. The electrons that are seen in association with the NADPH are not utilized in the transfer to molecular oxygen. This is in stark contrast to the use of the electrons in NAD that are used for the transfer to molecular oxygen. These electrons are destined for utilization in the reductive biosynthesis of compounds like fatty acids or steroids. The use of NADPH is seen in the reduction of glutathione using glutathione reductase and NADPH for the reducing electrons. NADPH is also imperative for the proper functioning of the cytochrome P-450 system and the process of phagocytosis.

Fig 24

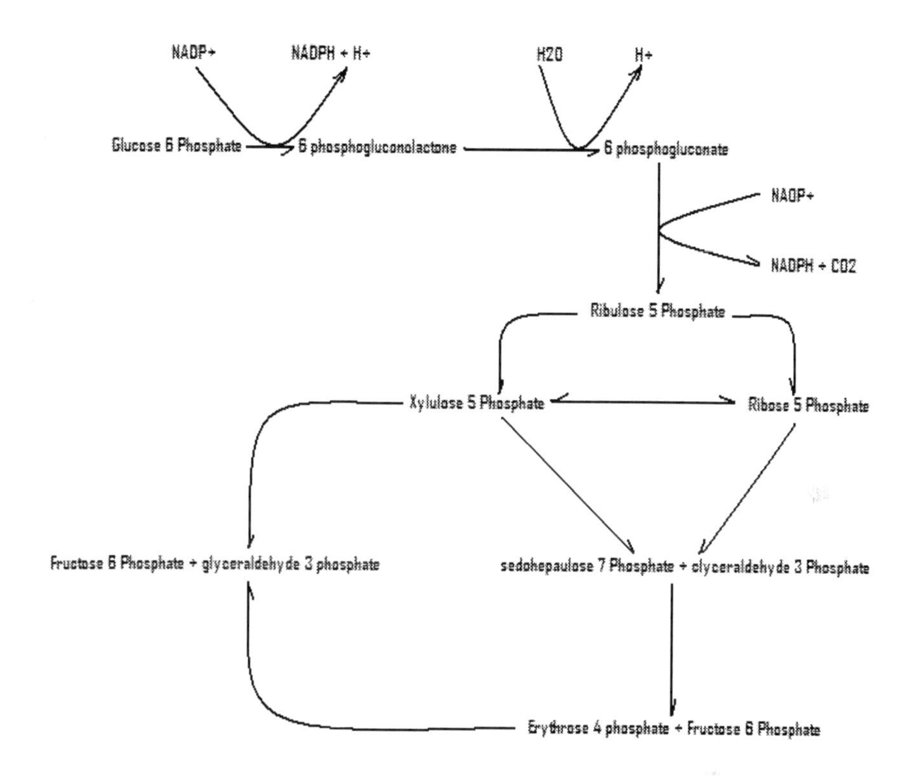
NADP+
NADPH + H+
H2O
H+
Glucose 6 Phosphate
6 phosphogluconolactone
6 phosphogluconate
NADP+
NADPH + CO2
Ribulose 5 Phosphate
Xylulose 5 Phosphate
Ribose 5 Phosphate
Fructose 6 Phosphate + glyceraldehyde 3 phosphate
sedohepaulose 7 Phosphate + clyceraldehyde 3 Phosphate
Erythrose 4 phosphate + Fructose 6 Phosphate

Phospholipid Metabolism

I. Phospholipids are polar, amphipathic molecules that have an alcohol bound by a phosphodiester bond to diacylglycerol or sphingosine. The hydrophilic head contains the phosphate while the hydrophobic tail is made from the 2 fatty acids. There are 2 classes of phospholipids.

   A. Phospholipids containing a glycerol backbone. These have a particularly essential role in the body as components of bile, components of surfactant and as membrane protein anchors. These are also known as Phosphoglycerides or glycerophospholipids.

      i. This subtype of phospholipids has phosphatidic acid attached to the carbon 3.

      ii. This is the simplest structure of the phospholipids.

      iii. This is also the precursor of the majority of the other phospholipids. The only cell unable to synthesize this compound is the erythrocyte.

Fig 25

```
                                   O
                                   ||
          O          CH2— O — C— RI
          ||          |
R2— C — O —— C-H           O
                      |            ||
                     CH2 — O — P — O-
                                   |
                                   O-
```

Phosphatidic Acid

iv. If the position at carbon 1 or carbon 3 has the fatty acid removed then the resulting structure is now a lysophosphoglyceride.

v. There is another important role for the Phosphoglycerides. This role is in the inner mitochondrial membrane as a compound called Cardiolipin. Cardiolipin is derived from the esterification of 2 of the phosphates to an extra glycerol molecule. This compound is important as it is the only phosphoglyceride that has antigenic activity.

Fig 26

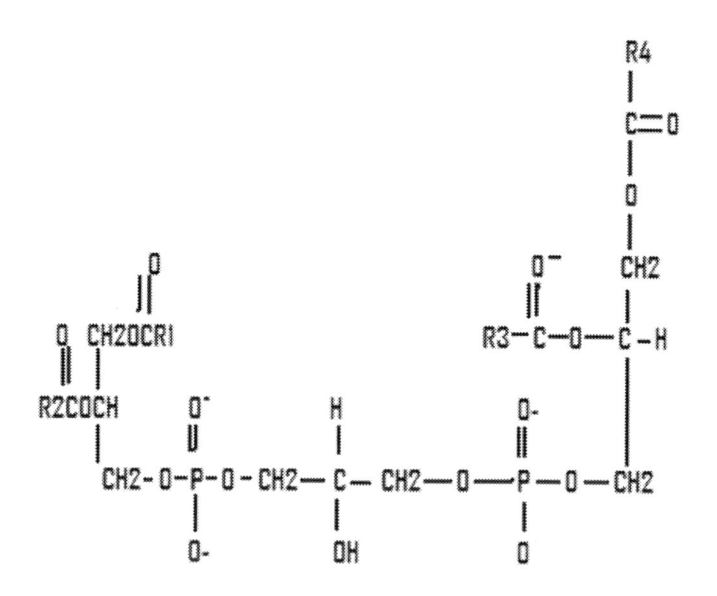

Cardiolipin

B. Those with sphingosine as the backbone are seen in membrane components.

i. If the compound has the backbone with an amino alcohol the compound is then termed a sphingosine.

ii. In this molecule, the attachment of the fatty acid to the amino group of the sphingosine through an amide linkage will yield a ceramide, another precursor of glycolipids. In this configuration, the alcohol at carbon 1 will be esterified to a phosphorylcholine. The product of this reaction will be a compound of sphingomyelin.

C. The synthesis of the phospholipids is done in the smooth endoplasmic reticulum for final processing in the Golgi apparatus.

i. The synthesis of this compound involves the transfer of phosphatidic acid from CDP diacylglycerol to an alcohol.

ii. Additional synthetic pathways are available from the transfer of the alcohol from the CDP alcohol mentioned above to 1,2 diacylglycerol.

iii. The synthesis of Phosphatidylethanolamine is accomplished through the phosphorylation of ethanolamine that is converted to an active form, CDP ethanolamine.

iv. The synthesis of phosphatidylcholine is performed in the same manner. In the lung, this compound will undergo transformation to become a component of surfactant. The conversion is in the type II granular cell of the lung. The end product of the reaction will yield diaplmytoylphosphatidylcholine. In this reaction the position 1 and position 2 of the glycerol molecule will be occupied by palmitate.

v. The synthesis of phosphatidylinositol will attach a molecule of steric acid to the carbon 1 of the glycerol molecule. This compound is important as it is a precursor for prostaglandin synthesis. It is also important in its role as an attachment for membrane proteins. Going back to the fluid mosaic model of the cell membrane, recall that the membrane has the ability to me motile. If there were attachment of the membrane proteins to other molecules, there would not be the fluidity as seen in the cell membrane. This lateral mobility is important for structural stability as it allows for the "movement" of those integral proteins. This attachment can be cleaved and it is important to have this cleavage ability as a second messenger system participant. Phospholipase C will act as the cleaving enzyme and the product of this enzymatic activity will be the liberation of diacylglycerol. The release of the diacylglycerol will activate the second messenger system.

D. Glycolipids are generally derived from ceremides as mentioned above. These compounds are found in the greatest concentration in the nervous tissue. Specifically, the ceremides are found in the outer layer of the plasma membranes and myelin sheaths.

i. Glycosphingolipids will differ from the sphingomyelin as they have no phosphate and the polar head is a monosaccharide or possibly a oligosaccharide that is directly attached to the ceramide through an O – glycosidic bond.

ii. Acidic glycosphingolipids carry a negative charge at physiologic pH. The negative charge is carried on the sulfate group of the Sulfatides or by the N-acetylneuraminic acid, NANA that is within the ganglioside.

Cholesterol Synthesis

I. Cholesterol is among the many compounds synthesized by many tissues within the human body. In humans, the tissues that are responsible for the majority of the cholesterol synthesis are the ovaries, testies, placenta, intestines, adrenal cortex and the liver. One of the most important functions of cholesterol in the human body is in the cell wall. It is responsible for the fluidity of the cell membrane. The concentration of cholesterol is directly related to the amount of structural stability the cell membrane will possess. Additional roles of cholesterol are found in the ability to synthesize bile acids steroid hormones. And additional roles in the production of Vitamin D.

Fig 27

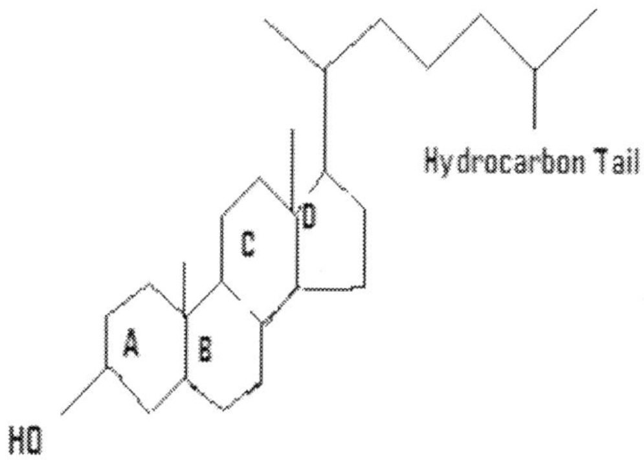

Hydrocarbon Tail

A B C D

HO

A. The structure of cholesterol is made from 4 rings that are fused together. Attached to the "D" ring is the hydrocarbon tail.

B. In the synthesis of cholesterol, the carbons are donated by acetate using NADPH for the reducing equalivents. The synthesis is found in the cytoplasm.

C. The initial reaction in the synthesis of cholesterol is the formation of 3 – hydroxy – 3 – methylglutaryl CoA (HMG CoA). This reaction is catalyzed by a thiolase in the formation of acetoacetyl CoA from the 2 molecules of acetyl CoA. For the formation of the end product there must be an additional acetyl CoA reacted to the acetoacetyl CoA that was just formed. For the additional acetyl CoA to be added in the next step of the reaction, the enzyme HMG CoA synthase will be required. This reaction will be in the endoplasmic reticulum. In this reaction there will be a liberation of a CoA molecule.

Fig 28

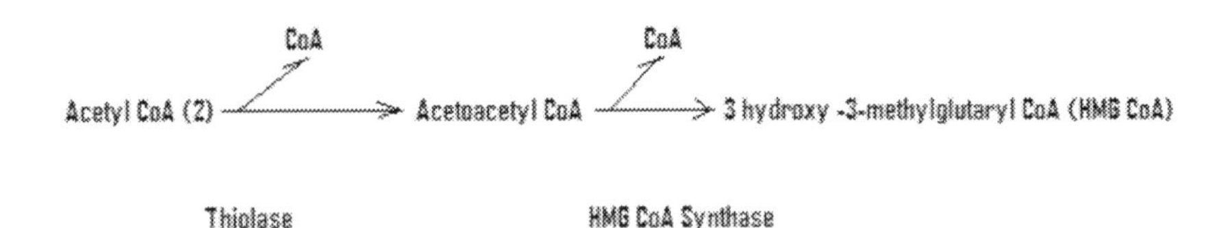

D. The following reactions are the final steps of the synthesis of cholesterol.

   i. Melvonic acid will react with kinases in a reaction that will require large amounts of ATP to produce 5 pyrophosphomevalonic acid.

ii. The next step involves the decarboxylation of 5-pyrophosphomevalonic acid to yield isopentenyl pyrophosphate (IPP). This step is energy expensive and will require the hydrolysis of ATP for the reaction to run.
iii. Next in the pathway will be an isomerization reaction of isopentenyl pyrophosphate to 3,3 dimethylallyl pyrophosphate (DPP) followed by the condensing of this molecule with IPP to yield geranyl pyrophosphate (GPP).
iv. Another condensation reaction is required as a second molecule of IPP will condense with GPP to yield farnsyl pyrophosphate (FPP).
v. The next step will be a combination and reduction reaction. This will be 2 molecules of farnsyl pyrophosphate releasing the pyrophosphate and then reduced to form squalene.
vi. Using molecular oxygen and NADPH in the hydroxylation of squalene will yield lanosterol. This is unique in the new molecule will be found as a cyclic structure rather than the planar position.
vii. The conversion of lanosterol to cholesterol is not a simple single step reaction. It will involve the reduction of the carbon backbone units from 30 units to 27 units and the cleavage of two methyl groups from C4. Additional reactions in this transformation will involve the movement of the double bond at C8 to C5 with the concurrent reduction of the double bond at the C24-25 position.

E. Metabolism of cholesterol is a difficult process as it is not a direct process. The sterol ring is eliminated directly through the excretion in the feces after being converted to a bile salt. The other mechanism of the metabolism of cholesterol is in the secretion of the cholesterol into the bile. In this step the cholesterol will be modified by the intestinal flora to coprostanol or cholestanol.

F. Bile is a mixture of phosphatidylcholine and bile salts. Common bile acids are cholic acids and chenodeoxycholic acid. The bile salts will be reused by the ability of the intestinal walls to reabsorb the salts. This is done in the ileum and requires energy investment as the process of reabsorption is via active transport. Once moved from the intestine the salts are sent to the portal circulation, then on to the liver for modification and then passed to the duodenum. In the duodenum, some of the salts will be converted to acids and returned to the liver through the enterohepatic circulation.

i. The structure of a bile acid contains 24 carbons with 2 or possibly 3 hydroxyl groups with a side chain that has termination in a carboxyl group.

ii. These bile acids are amphipathic with the methyl groups lying in the β position and the hydroxyl groups occupying the α position.

iii. The orientations of the methyl and the hydroxyl groups will provide for a polar and nonpolar face. This arrangement makes the acid a tremendous emulsifying agent. This emulsification prepares the triacylglycerol and other lipids for action by the pancreatic enzymes.

iv. In the liver, the majority of the synthesis of the bile acids will be completed. There will be the addition of the double bond at the B ring of the cholesterol molecule and the insertion of hydroxyl groups with concurrent shortening of the hydrocarbon chain by 3 carbons with the final modification being the carboxyl group attached to the end of the chain. The acid must then be conjugated to either glycine or taurine with an amide bond between the carboxyl of the bile acid and the amino of the compound to be bound.

v. Once bound the structure is now a bile salt and the naming of the compound is glycocholic or glycochenodeoxycholic acid and taurocholic or taurochenodeoxycholic acids. It is at this point the bile salts can be found in the bile.

vi. The removal of the amino glycine and the taurine will be through the action of the intestinal bacterial flora. The end product of this series of reactions will result in the formation of secondary bile acids, deoxycholic and lithocholic acids.

G. Lipoproteins or apolipoproteins are complexes of lipids with associated proteins. The lipoproteins consist of Chylomicrons (CM), very low density lipoproteins (VLDL), low density lipoproteins (LDL), and high density lipoproteins (HDL). These are the packaging and delivery forms of the lipids to the tissue. The lipids that are carried are typically triacylglycerols and cholesterol. Structurally, the lipoproteins are composed of a neutral core of either triacylglycerol or cholesteryl esters or a combination of both. The surrounding shell is composed of apolipoproteins, phospholipids, and a nonesterified cholesterol. The orientation of these outer shell components is very important as the polar aspect of the molecules must be available for interaction with the aqueous outer environment.

Fig 29

Percent composition	Very Low Density Lipoprotein	Low Density Lipoprotein	High Density Lipoprotein	Chylomicron
Triacylglycerol	60	8	5	90
Protein	5	20	40	2
Phospholipid	15	22	30	3
Cholesterol & Cholesteryl Esters	20	50	25	5

H. Metabolism of the chylomicrons that are responsible for the transport of the dietary triacylglycerols, cholesterol, and cholesteryl esters to the tissues is enzymatically driven.

i. The chylomicrons are synthesized as they are moved along from the endoplasmic reticulum to the golgi for packing with the TAG's, cholesterol, and phospholipids. The initial synthesis is begun on the rough endoplasmic reticulum. The chylomicrons are packed into secretory vessicles for release and transport in the lymphatic system.

ii. The chylomicron as it is released is called a nascent chylomicrons since it contains apolipoprotein B-48, but it is modified in the plasma through the addition of apoE and C apolipoproteins. The apo C is important since it is a Isoenzyme of this apo C, apo CII that will stimulate the activitation of lipoprotein lipase that will degrade the TAG within the chylomicron.

iii. After the degradation of the chylomicrons TAG component, the size of the chylomicrons is greatly reduced. There is an inverse relation between the size and the density of the chylomicrons. This altered chylomicron is removed from the circulation by the liver as the hepatocytes recognize and bind the apo B-48 and the apo E proteins. They will become bound to the lysosome for hydrolytic degradation.

I. Metabolism of very low density lipoproteins that were produced in the liver is through the lipoprotein lipase acting on the same TAG as above.

i. VLDL's are released from the liver just as the chylomicrons. The difference is in the lipoprotein component apo B-100 and apo A-1. The VLDL must obtain the apo C and the apo E from the HDL that is circulating in the blood.

ii. The VLDL in the circulation is acted upon just as the chylomicron, having the TAG removed through the action of the apo CII activated lipoprotein lipase and producing a smaller more dense molecule. In this processing of the VLDL, there is a transfer of the apo C and apo E from the VLDL to the HDL with the cholesteryl ester being transferred from the HDL to the VLDL and sending the TAG or the phospholipid from the VLDL to the HDL. Once these modifications are completed the VLDL is converted to IDL in the plasma.

J. The metabolism of the low density lipoproteins is achieved through the recognition of the surface apo B-100. The low density lipoprotein has much less triacylglycerol than VLDL precursors.

i. The main function of the LDL is the provision of cholesterol to the periphery. The receptor is negatively charged and are found within the clathrin coated pits on the cell membrane.

ii. Upon binding, the LDL in endocytosed and they have the clathrin coat digested and multiple LDL molecules will conjugate to form an endosome.

iii. The endosome will have the internal pH reduced and thus forcing the LDL to the center of the vessicle and the receptors to the other side. The enzyme that is responsible for the pH differential is the endosomal ATPase.

K. The cellular cholesterol levels are affected by the amounts of circulating chylomicron remnants, HLD and LDL components. The levels of these components will either stimulate the action of HMG

CoA reductase. If the levels are low then there is an additional enzymatic activity to produce a form that is available for storage. Esterified processing is through the action of acyl CoA:cholesterol acyltransferase (ACAT) to transfer the fatty acid from acyl CoA to the cholesterol. The reaction of these two substrates will yield a molecule of cholesteryl ester. The cholesteryl ester is the storage form of the cholesterol. Activity of the synthesis mechanism of the LDL receptors will be inhibited if the transcription of the gene is limited.

L. Synthesis of the high density lipoproteins begins in the liver and subsequent release into the blood stream as is the case of the previously mentioned lipoproteins. The function of the high density lipoproteins is that of being a storage unit for apoC II and transporting the cholesteryl esters to the liver.

M. The esterification of the free cholesterol is done as the free cholesterol is uptaken by the HDL. There is activity of the enzyme phosphatidylcholine: cholesterol acyltransferase (PCAT) that is responsible for the transfer of the cholesteryl ester to the LDL from the VLDL. In this process, the direct transfer of the carbon from carbon 2 of the phosphatidylcholine to the cholesterol. This reaction will result in the formation of lysophosphatidylcholine. The result of this reaction on the cholesteryl ester will be effectively sequestered by the hydrophobicity of the ester. For the removal of HDL there must be the transfer of the HDL to the VLDL. More than 2/3 of the cholesterol of the plasma is esterified to a fatty acid.

II. Cholesterol is well known to be the precursor for the classes of steroid hormones manufactured by the body. The classes of steroid hormones include: glucocorticoids, mineralcorticoids, and sex hormones. The sex hormones are further subdivided into the androgens, the estrogens and the prostaglandins. The structure of the hormones are such that they require the carrier proteins. These carrier proteins, like transcortin responsible for the transport of cortisol and cortisterone or sex hormone binding protein that is responsible for the transport of sex proteins.

A. The synthesis of the sex hormones are derived from the shortening of the cholesterol and the concurrent hydroxylation of the nucleus of the steroid. This reaction is catalyzed by the enzyme desmolase. The action of this enzyme is the conversion of cholesterol to pregnenolone. This step in the reaction pathway utilizes the cofactor NADPH and molecular oxygen.

i. Pregnenolone will be oxidized and isomerized to progesterone.

ii. Progesterone will be hydroxylated to the other steroid hormones.

Fig 30

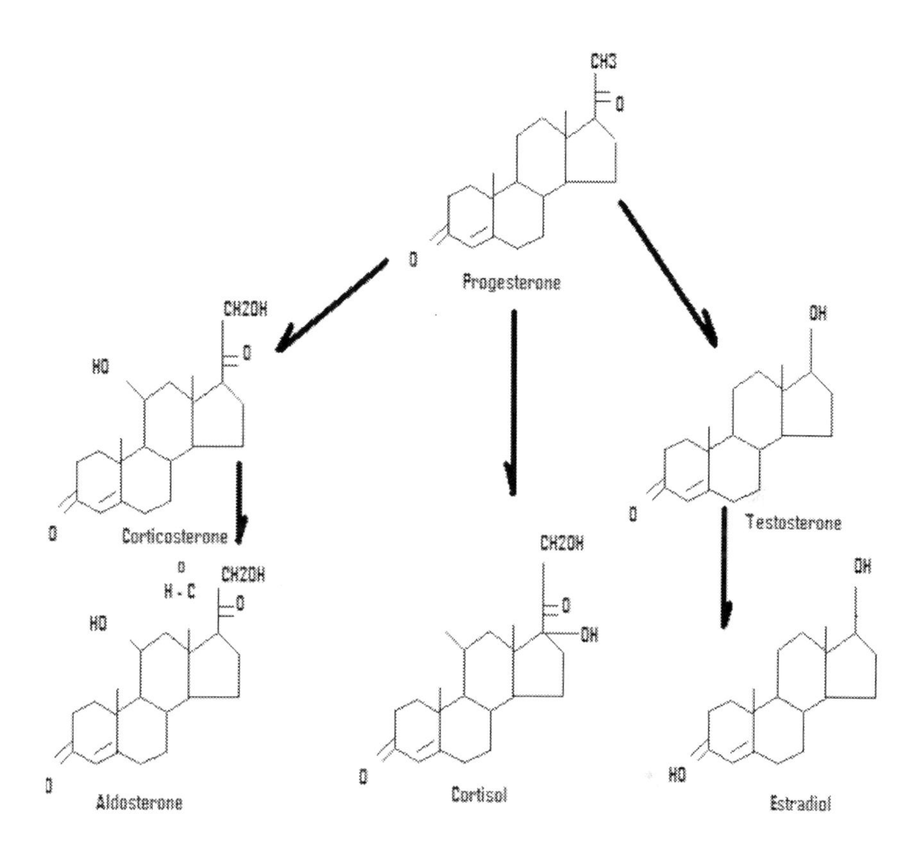
CH3
O
O
Progesterone
CH2OH
O
HO
O
Corticosterone
O
H - C
CH2OH
O
HO
O
Aldosterone
CH2OH
O
OH
O
Cortisol
OH
O
Testosterone
OH
HO
Estradiol

Metabolism Of Amino Acid Carbons

I. All amino acids are either glucogenic or ketogenic. The defining aspect of this is the products of the metabolism of the amino acid into its components.

Fig 31

AMINO ACIDS	**GLUCOGENIC AMINO ACIDS**	**KETOGENIC AMINO ACIDS**	**BOTH KETOGENIC & GLUCOGENIC AMINO ACIDS**
NONESSENTIAL AMINO ACIDS	ALANINE ASPARGININE ASPARTATE CYSTEINE GLUTAMATE GLUTAMINE GLYCINE PROLINE SERINE		TYROSINE
ESSENTIAL AMINO ACIDS	ARGININE HISTIDINE METHIONINE THREONINE VALINE	LEUCINE LYSINE	ISOLEUCINE PHENYLALANINE TRYPTOPHAN

A. Glucogenic amino acids will yield pyruvate of an intermediate of the TCA cycle as they are metabolized to end product. They are able to produce glucose as the end result of the gluconeogenic pathway.

i. Oxaloacetate is an intermediate of the Citric acid cycle (TCA) or (Kreb's cycle). Formation of this intermediate is from the hydrolysis of asparagine that will yield the amino acid aspartate and leaving an ammonia molecule (NH_3).

ii. Alpha ketogluterate is also an intermediate of the Citric acid cycle. There are several amino acids that are able to be reacted upon to generate this pathway intermediate. Glutamine is going to be reacted upon to produce glutamate. The enzyme that is responsible for this reaction is glutaminase. In a second part of the conversion, there is a transamination reaction that will produce alpha ketogluterate from the glutamate left as the product of the previous step. There is an additional path that glutamate can be reacted on to produce the alpha ketogluterate. In this additional method, there is an oxidative deamination of glutamate that will elicit the alpha ketogluterate that is the intermediate of the TCA.

iii. Arginine will generate alpha ketogluterate, but not directly. Arginine will require conversion to ornithine. The enzyme responsible for the catalysis of this reaction is arginase. A second enzyme will catalyze a second reaction of transamination. This is the transamination of ornithine that will yield glutamate γ semialdehyde. This is still not the end of the path for the conversion of arginine to alpha ketogluterate. In the final step of this path, there is a conversion of glutamate γ semialdehyde to alpha ketogluterate.

iv. Alpha ketogluterate is also produced from the oxidation of proline to Δ^1 –pyrroline 5 carboxylate. This product will be reacted upon again through a second oxidation reaction to yield glutamate. Once this step is completed, the remaining conversion of glutamate to alpha ketogluterate is just as in each other reaction. It will be transaminated or oxidatively deaminated.

v. Alpha ketogluterate is also formed from the amino acid histidine. This is again a multi step reaction in the conversion to the TCA intermediate. Histidine will be deaminated first and then hydrolyzed, producing N forminoglutamate. This product cannot alone be used as a step for the production of alpha ketogluterate, so it will act as the donor for the formimino group to the acceptor Tetrahydrofolate. At this point the reaction will leave Tetrahydrofolate that will be degraded at a point in the future.

vi. Pyruvate is also an intermediate of the TCA cycle, however not directly.

a. Alanine will be acted upon by the enzyme Alanine aminotransferase to cleave the amino group in this transamination reaction.

b. Serine dehydratase is the enzyme responsible for the conversion of serine to pyruvate. If required, serine can be converted to glycine first and N^5 N^{10} methylenetetrahydrofolic acid.

Glycine can be converted to serine if there is an addition of the methylene group from the N^5 N^{10} methylenetetrahydrofolic acid. If it is not to be converted to serine in this pathway it will be oxidized to CO_2 and NH_4^+.

c. Cysteine will loose its sulfur in a desulfuration reaction after it has been reduced from cystine. This reduction reaction is as all other reduction reactions in need of a reducing agent or electron carrier. In this reaction that role is filled by NADH + H^+.

d. Theonine is unusual in the additional intermediate that it can become on its way to become pyruvate. Threonine will become a precursor for succinyl CoA as it will metabolize to alpha ketobutyrate.

vii. There are also amino acids that can be metabolized to a third intermediate of the TCA cycle. Fumarate is an intermediate of the TCA cycle and it can be formed from tyrosine and phenylalanine. This is also unusual in that phenylalanine and tyrosine are KETOGENIC & GLUCOGENIC. If phenylalanine is hydroxylated it will yield tyrosine. The enzyme responsible for this is phenylalanine hydroxylase. This reaction also requires an additional cofactor, tetrahydrobiopterin + O_2.

viii. Succinyl CoA is the intermediate of the TCA cycle that is derived from the metabolism of methionine, valine, Isoleucine, and Threonine.

B. Ketogenic amino acids will yield one of three end products. Acetoacetate, Acetyl CoA and acetoacetyl CoA are the three end products that will be derived from the ketogenic amino acids.

i. Leucine is reacted upon by the enzyme, branched chain α amino acid transferase to yield the product α ketoisocaproic acid. Additional enzymatic activity of branched chain α keto acid dehydrogenase will yield Isovaleryl CoA. Cofactors for this reaction are TPP, CoA, Lipoic acid, NAD, FAD. This reaction is an oxidative decarboxylation. The end product of this reaction will be acetoacetyl CoA, after a dehydrogenation of Isovaleryl CoA.

ii. Lysine is a special case of the metabolism of the amino acids to yield acetoacetyl CoA. It is special because it requires conversion to a unusual intermediate, α aminoadipate – δ semialdehyde. Once this intermediate is formed, the conversion will continue to the final product of acetoacetyl CoA.

iii. Tryptophan is also a converted to acetoacetyl CoA.

C. Branched chain amino acids are utilized in the production of these high energy compounds through the reactions mentioned above. The use of these amino acids for synthesis of the high energy compounds is done in the peripheral tissue and not the liver.

Fig 32

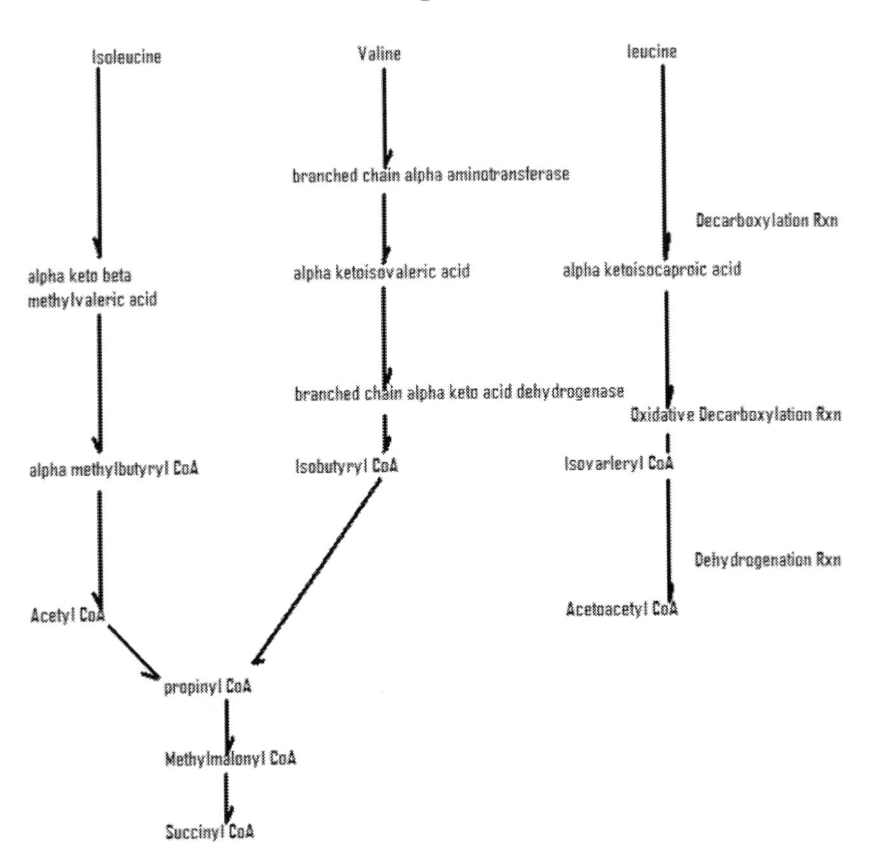

II. Synthesis of some important amino acids will be required as the nonessential amino acids that are needed for normal development of many tissues. The greatest amount of nitrogen in the world is found bound to amino acids and nucleotides. In able to cleave the bond of nitrogen, N_2, a triple bond that is very stable, there is a tremendous amount of energy that will be required ($\Delta G_o' = 945$ kJ/mol). The enzyme that is responsible for this cleavage is nitrogenase and it requires a high energy donor from ATP with the appropriate reducing equivalents. The form of the nitrogen that is circulating is in the form of ammonia. This ammonia is incorporated into the amino acid glutamate ammonia ligase and also from the enzyme glutamate dehydrogenase. These reactions are found within all living beings. This reaction is also capable of generating other amino acids through transamination reactions.

A. Glutamine is synthesized from the amino acid glutamate. For this conversion to take place the enzyme required to catalyze this reaction is glutamine synthase. This is a very energy expensive amidation reaction to run and therefore uses ATP to drive the reaction. There is an additional function of this reaction. It is important as a detoxification of the brain and liver for excessive ammonia that is not cleared in the urea reaction pathway.

B. Asparginine is formed in a like manner as that described with glutamine. In this reaction pathway, aspartate will be reacted upon by the enzyme asparginine synthase. This is another example of a very energy expensive reaction in that it will also require the hydrolysis of ATP to be driven. In this reaction the amide group will be donated from glutamine.

C. The synthesis of proline will also utilize glutamate as an intermediate in the synthetic pathway. In this pathway the action of the enzyme glutamate γ semialdehyde will yield Δ^1 pyrroline 5 carboxylate. The product of this reaction will then be reduced and the product of this final reaction of the conversion will produce proline.

D. Glycine is synthesized in the simple reaction of the removal of a hydroxymethyl group from serine.

E. Serine is synthesized in a series of reactions that begin with an intermediate of the glycolysis pathway. This intermediate is glyceraldehyde 3 phosphate. In the formation of the final product of serine, the glyceraldehyde 3 phosphate will first be oxidized to 3 phosphopyruvate. Once this reaction is run, there will be a transamination of the product to yield 3 phosphoserine that will undergo a hydrolysis of the phosphate.

F. Cysteine is synthesized in a series of reactions. The first of the reactions will be the conversion of methionine to S –adenosylmethionine by the action of the enzyme S adenosylmethionine synthase. This is a very energy expensive reaction that has ATP hydrolyzed and utilizes Mg^{2+} as a cofactor. The next reaction will involve the enzyme methyltransferase that will catalyze the conversion of S adenosylmethionine to S adenosylhomocysteine. The path to Cysteine is continued as the adenosine will be removed yielding homocysteine. Homocysteine will in turn be reacted upon by cystathionine synthase. The product of this reaction will be Cystathionine. In the next and final step of the reaction, there is the reaction of the enzyme γ cystathionase on cystathionine that will produce cysteine.

E. Tyrosine is an important amino acid in that it is a precursor of Dopa, a neurotransmitter. Synthesis of tyrosine will be from the amino acid phenylalanine.

In order to have this conversion run, the enzyme phenylalanine hydroxylase is required as arc the cofactors, tetrahydrobiopterin and molecular oxygen.

F. Essential amino acids are Histidine, isoleucine, leucine, lysine, methionine, phenylalanine, threonine, tryptophan and valine.

G. Non-essential amino acids are

a. Alanine, arginine, asparginine, aspartate, cysteine, glutamate, glutamine, glycine, proline, serine, tyrosine.

IV. In the scope of the synthesis of the amino acids Glutamate, glutamine, alanine, aspartate, asparginine will be generated from the pool of amino acids that are able to be utilized in the gluconeogenic pathway. In humans, glutamate is the central point of entry of amino acids.

The reaction is: Glutamate + ATP + NH_3 = glutamine + ADP + Pi. This reaction is found in the mitochondria of all mammals.

Programmed Cell Death

I. Cell growth and regulation of division is a very complex process. Division of this process is regulated by the cyclin-dependentkinases (Cdks). The activity of this enzyme is to phosphorylation of serine and or threonine side chains. The regulation of the enzyme is done via protein kinases and phosphatases. There is also a requirement for the inhibitor of the Cdk to be activated.

A. The growth phases are divided into S and M phases. The S phase is where we see duplication of the chromosomal DNA.

B. M phase is where we see mitosis, yielding the 2 progeny.

1. DNA condenses
2. Centrosomes develop the mitotic spindles
3. Degradation of the nuclear membrane
4. The spindle will connect the chromosomes
5. The chromosomes will align along the equator of the cell.
6. All chromatids will separate to yield 2 sets of chromosomes
7. Chromosomes will form individual nuclear chromosomes.
8. Condensation of the chromosomes.
9. Cytokinesis

C. Separation of the S and the M phase are the Gap phases G1 G2 G0(G phases).

D. The M phase is shorter than the other phases but is more complex than the others. In human replication, the M phase (Cdk or Cdk-1) is known to bind to M cyclin (cyclin B) to form the heterodimer that is the probgation for the path. This complex formation is required for the continuation of the apoptotic process.

i) for the cell to enter into the M phase is done via the phosphorylation of M-Cdk/M -cyclin by CAK and concurrent removal of the inhibitory phosphate by Cdc-25 phosphatase.

ii) The M -cyclin used will only be synthesized during the G_2 phase.

II. Apoptosis is the process of cell death. In the instance of the description the apoptosis is different from that of necrosis. There is a very specific path of difference between these two types of cell death. Apoptosis is a programmed cell death while the necrotic cell death is the result of the process of an inflammatory reaction. In many instances the apoptosis is not observed until after the event occurs. Characteristics of apoptosis are the changes in the cytoskeleton, nuclear DNA and the plasma membrane itself. The signals of the apoptotic generators will initiate a response to activate the caspases and induce the initiation of apoptosis. Caspases are cytoplasmic protease enzymes that are designed to hydrolyze specific peptide bonds. The hydrolysis of the peptide bond will increase or decrease the enzymatic activity. Endonucleases will degrade the DNA that yield segments of approximately 170 base pairs in size. This will allow for the remaining portions of the cell to be available for attack by the macrophages for phaygocytosis.

A. factors that induce the apoptosis are damaged DNA

B. Improper entry into the S phase

C. Inadequate growth factors.

D. Presence of death signal proteins

III. The manner in which the cell carries out the process of apoptosis is mediated through two pathways of action, the death receptor pathway and the mitochondrial pathway. Each pathway will lead to the same outcome, that of cellular death, but the manner each path leads to the destruction is unique.

A. Mitochondrial pathway is considered to be an intrinsic pathway. This pathway is regulated by the activity of the Bcl 2 like protein. This protein is found within the outer mitochondrial membrane, and hence the term intrinsic. The function of the Bcl 2 like protein is to regulate the translocation of specific substances from the mitochondrial inter- membrane space into the cytosol of the cell.

B. This specific protein, the Bcl 2 like protein has 24 sub classifications of it. The main divisions are those that have been identified as having anti - apoptotic properties and those that have pro - apoptotic properties. There are additional pro- apoptotic facilitator proteins that have been identified as well.

i) The pro-apoptotic proteins will create pores using the proteins Bax and Bak. These proteins express the domains of BH1, BH2 and BH3 within the structural make up.

ii) The anti- apoptotic proteins are identified as Bcl 2 and BcL $X_{L.}$ The Bcl proteins share a common characteristic of the BH3 domain. However, the Bcl 2 and the BcL X_L proteins are seen to express four domains in common, the BH1, BH2, BH3, and the BH4 domains.

iii) The facilitator proteins are identified as Bid, Bad, PUMA, Noxa proteins. These facilitator proteins will express only the BH3 domain.

C. These proteins share the common interest of altering the flux of cytochrome c through the mitochondrial membrane. The mechanism of this flux is based on the protein subdivision that is expressed. Bax and Bak will form pores that will allow increased flow while the Bcl 2 and the Bax proteins will prevent the flux of cytochrome c. The mechanism of action is the manner it competitively binds to the Bak and the Bax proteins. By binding in the way that they do the binding prevents the formation of the pores they are designed to create. Preventing the pore formation will prevent the flow of cytochrome c.

D. The Death receptor path is considered to be an extrinsic pathway as it requires the ligand in the extra cellular side of the cell membrane to bind and activate the path. This pathway is analogous to the second messenger system in the manner the binding of the receptor on the extra cellular aspect of the membrane will induce intracellular modifications that will induce cellular death. In the cell the receptor is the FADD receptor located on the intracellular side of the membrane. The death inducing signaling complex will then initiate the activity of the death domain. The death domain (DD) is found to bind with the receptor for the pro-caspase 8 that will be hydrolyzed in the path to the active form of the caspase.

E. The balance between the subunits of the protein is the deciding factor for the activity of the apoptosis pathways. In the case of the mitochondrial path, the increased levels of the Bax and the Bak proteins are known to prevent the loss of cytochrome c and thus cell apoptosis. So it is clear that the balance between the pro apoptotic and the anti-apoptotic proteins hold the key to the activity of the pathways. There is also some additional protein bases other than cytochrome c involved in the pathway. On the outside of the mitochondria Apafl (Apoptotic protease activating factor 1) is one example of these additional factors involved in the pathway. In this protein - receptor complex design this subunit is termed an apoptosome. It is this subunit that is responsible for the activation of pro-caspase 9 to caspase 9. In the same setting there are additional proteins that can begin the sequence of events that result in the apoptosis of the cell. These additional proteins are the Smac - DIABLO are able to induce the mitochondrial pathway to initiate apoptosis.

F. p53 induced Apoptosis is one of the most well studied of the apoptosis mechanisms. In this pathway there is DNA damage that is recognized by the Mdm2 protein that

G. MAPK pathways are unique pathway in the manner is has regulatory effects in both cell division and cell apoptosis. The

i) Growth factor receptor signals are transmitted to the GTP -Ras protein that in turn will activate the cascade MAPKKK→ MAPKK → MAPK path. Since there are multiple MAPK paths

ii) Since there are multiple MAPK paths, it is possible to have other methods of activation other than the Ras activation such as UV light, radiation, cytokines, and inflammatory pathway stimulation.

iii) One of the specific MAPK paths also known as JNK is well known to participate in apoptosis through the phosphorylation of the Bcl-2 protein. This phosphorylation elicits increased mitochondrial concentration of Bcl and Bax proteins thus producing increase rate of apoptosis. This JNK path is also known to phosphorylate p53 oncogene and promote apoptosis.

H. Weel 1 is considered to be a G2/Mphase kinase. The function of this enzyme is the phosphorylation of M-Cdk at the inhibitory site preventing the M-cyclin /M-Cdk activity and entry into the M phase. There are at least 4 different G1/S-Cdk/Cyclin complexes that will regulate the cell as it progresses through G1 and the S phases. These are directed by the CAK in conjunction with a specific Weel analogue kinase. Within this segment of reaction steps, there are phosphatases that are required to cleave the inhibitory phosphates and thus allow the activity of the protein to come to action. These G1 and Cdk proteins are also required in various phases of the cell cycle outside those mentioned above. Some of the Cdk's are required in late G1 phase so that the cell may proceed into the S phase. Without these Cdk's this progression into S phase would not occur. There are come Cdk's that are required for the progression of the cell from S phase into the G2 phase. Ubiquine will be used in the regulation of the G1/S Phase through its activity on the Cdk's present. The system of this ubiquination system is referred to as the SCF complex.

I. The p53 regulators are the transcription factors associated with damaged DNA. The DNA is monitored by specific kinases as they scan the strand and upon recognition of a defect, they will become activated and increase the synthesis of the p53 itself.

Fig 1

1 1A	2 2A	3	4	5	6	7	8	9	10	11	12	13 3A	14 4A	15 5A	16 6A	17 7A	18 8A
1 H 1.008																	2 He 4.003
3 Li 6.941	4 Be 9.012											5 B 10.81	6 C 12.01	7 N 14.01	8 O 16.00	9 F 19.00	10 Ne 20.18
11 Na 22.99	12 Mg 24.31											13 Al 26.98	14 Si 28.09	15 P 30.97	16 S 32.07	17 Cl 35.45	18 Ar 39.95
19 K 39.10	20 Ca 40.08	21 Sc 44.96	22 Ti 47.88	23 V 50.94	24 Cr 52.00	25 Mn 54.94	26 Fe 55.85	27 Co 58.93	28 Ni 58.69	29 Cu 63.55	30 Zn 65.38	31 Ga 69.72	32 Ge 72.59	33 As 74.92	34 Se 78.96	35 Br 79.90	36 Kr 83.80
37 Rb 85.47	38 Sr 87.62	39 Y 88.91	40 Zr 91.22	41 Nb 92.91	42 Mo 95.94	43 Tc (98)	44 Ru 101.1	45 Rh 102.9	46 Pd 106.4	47 Ag 107.9	48 Cd 112.4	49 In 114.8	50 Sn 118.7	51 Sb 121.8	52 Te 127.6	53 I 126.9	54 Xe 131.3
55 Cs 132.9	56 Ba 137.3	57 La* 138.9	72 Hf 178.5	73 Ta 180.9	74 W 183.9	75 Re 186.2	76 Os 190.2	77 Ir 192.2	78 Pt 195.1	79 Au 197.0	80 Hg 200.6	81 Tl 204.4	82 Pb 207.2	83 Bi 209.0	84 Po (209)	85 At (210)	86 Rn (222)
87 Fr (223)	88 Ra 226	89 Ac† (227)	104 Unq	105 Unp	106 Unh	107 Uns	108 Uno	109 Une									

*Lanthanides	58 Ce 140.1	59 Pr 140.9	60 Nd 144.2	61 Pm (145)	62 Sm 150.4	63 Eu 152.0	64 Gd 157.3	65 Tb 158.9	66 Dy 162.5	67 Ho 164.9	68 Er 167.3	69 Tm 168.9	70 Yb 173.0	71 Lu 175.0
†Actinides	90 Th 232.0	91 Pa (231)	92 U 238.0	93 Np (237)	94 Pu (244)	95 Am (243)	96 Cm (247)	97 Bk (247)	98 Cf (251)	99 Es (252)	100 Fm (257)	101 Md (258)	102 No (259)	103 Lr (260)

Fig 2

Number of 3333 Bonds	Number of unused e- Pairs	Hybrid Orbital	Bond Angle	Geometry	Example
2	0	sp	180	Linear	BeF2
3	0	sp^2	120	Trigonal Planar	BF3
4	0	sp^3	109.5	Tetrahedral	CH4
3	1	sp^3	90 - 109.5	Pyramidal	NH3
2	2	sp^3	90- 109.5	angular	H2O
6	0	$sp^3 d^2$	90	Octahedral	SF6

Fig 3

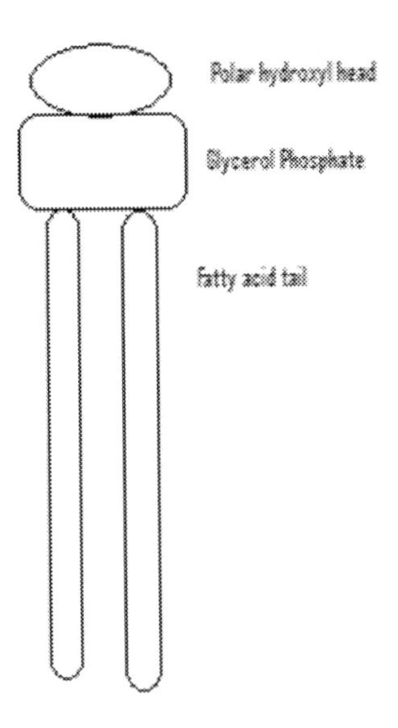

Fig 4.

Organelle	***Function***
Plasma Membrane	Transport of ions and molecules Recognition Receptors for small and large molecules Cell morphology and movement
Nucleus	DNA synthesis RNA synthesis
Nucleolus	RNA synthesis and ribosome synthesis
Endoplasmic Reticulum	Membrane synthesis Synthesis of proteins and lipids for export Export proteins Detoxification reactions
Golgi Apparatus	Protein exportation Post-translational protein modification
Mitochondria	Cellular respiration Synthesis of Urea and heme Oxidation of lipids and proteins Energy Production
Microtubules / Microfilaments	Cellular motility and mobility Cytokeletal structure
Lysosome	Cellular debris clearing Hydrolysis of lipids, proteins, carbohydrates and nucleic acids
Perioxomes	Oxidative reactions of cellular byproducts
Cytosol	Metabolic reactions involving the carbohydrates, amino acids, nucleotides while synthesizing proteins

Fig 5

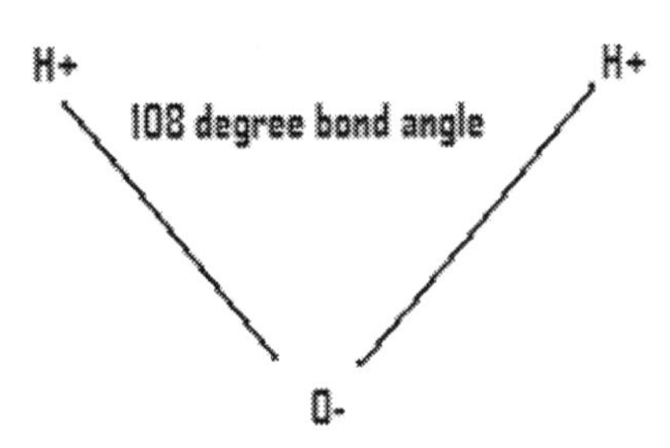

Fig 6

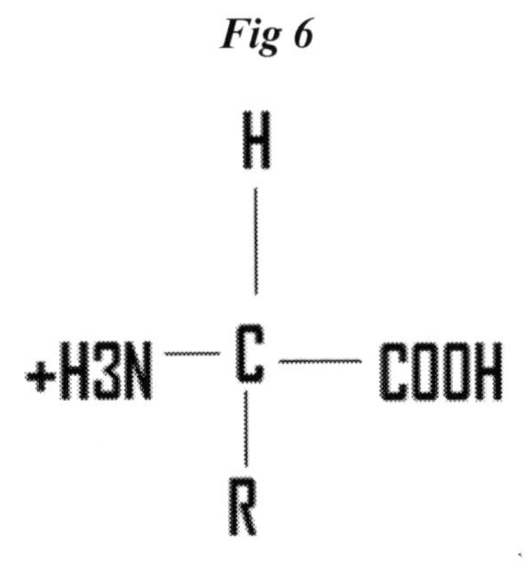

Fig 7

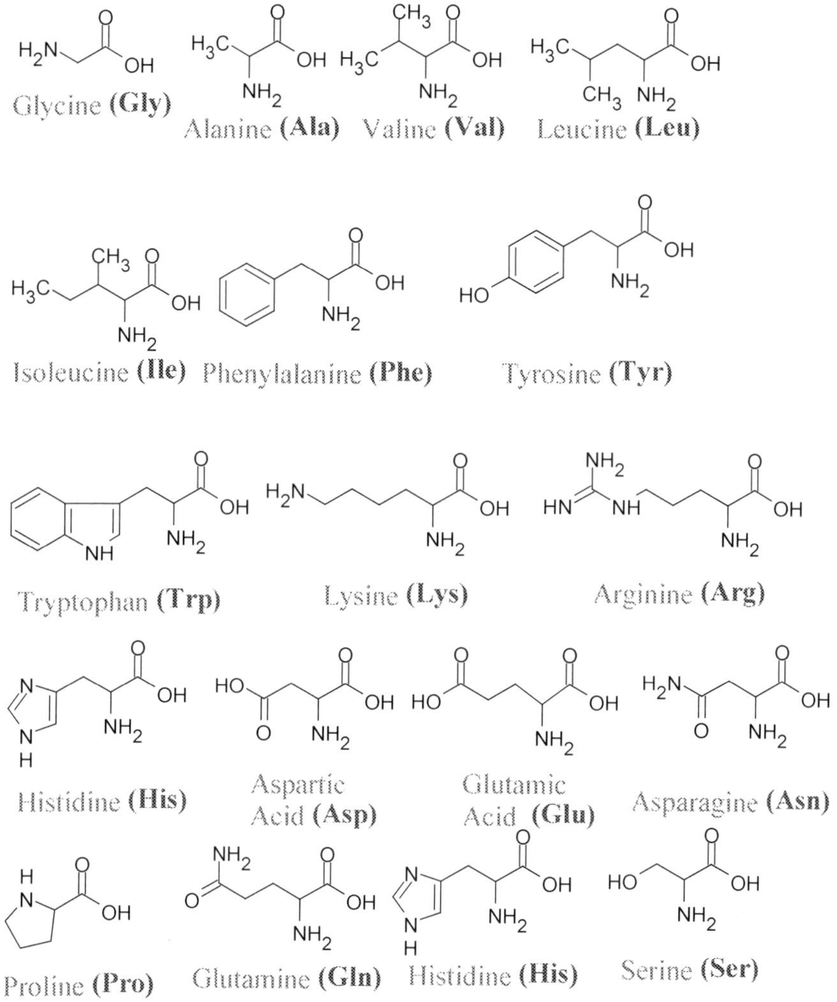

Glycine **(Gly)** Alanine **(Ala)** Valine **(Val)** Leucine **(Leu)**

Isoleucine **(Ile)** Phenylalanine **(Phe)** Tyrosine **(Tyr)**

Tryptophan **(Trp)** Lysine **(Lys)** Arginine **(Arg)**

Histidine **(His)** Aspartic Acid **(Asp)** Glutamic Acid **(Glu)** Asparagine **(Asn)**

Proline **(Pro)** Glutamine **(Gln)** Histidine **(His)** Serine **(Ser)**

Fig 8

COMPOUND	PRECURSOR	PRIMARY FUNCTION
Dopamine	Tyrosine	Neurotransmitter
Epinepherine	Tyrosine	Hormone
GABA	Glutamate	Neurotransmitter
Histamine	Histidine	Vasodilator
Melanin	Tyrosine	Pigment
Melatonin	Tryptophan	Hormone
Norepinepherine	Tyrosine	Neurotransmitter
Serotonin	Tryptophan	Vasoconstrictor
Thyroxine	Tyrosine	Hormone

Fig 9

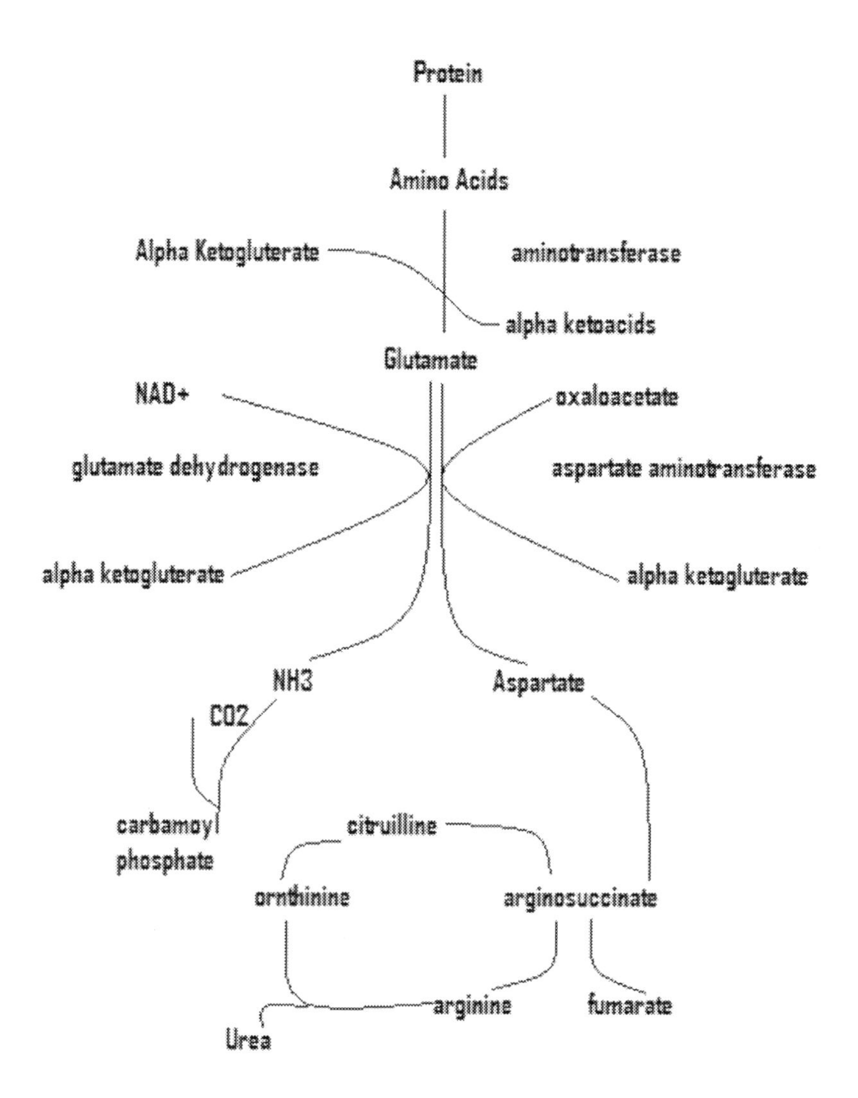
Protein
Amino Acids
Alpha Ketogluterate
aminotransferase
alpha ketoacids
Glutamate
NAD+
oxaloacetate
glutamate dehydrogenase
aspartate aminotransferase
alpha ketogluterate
alpha ketogluterate
NH3
Aspartate
CO2
carbamoyl
phosphate
citruilline
ornthinine
arginosuccinate
arginine
fumarate
Urea

Fig 10

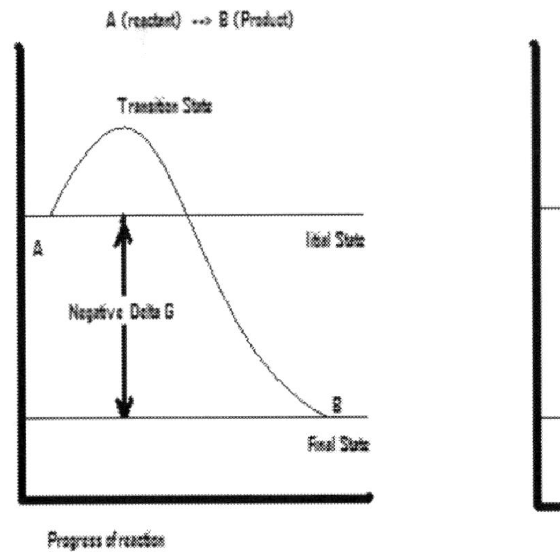
A (reactant) --> B (Product)
Transition State
A
Negative Delta G
B
Final State
Progress of reaction

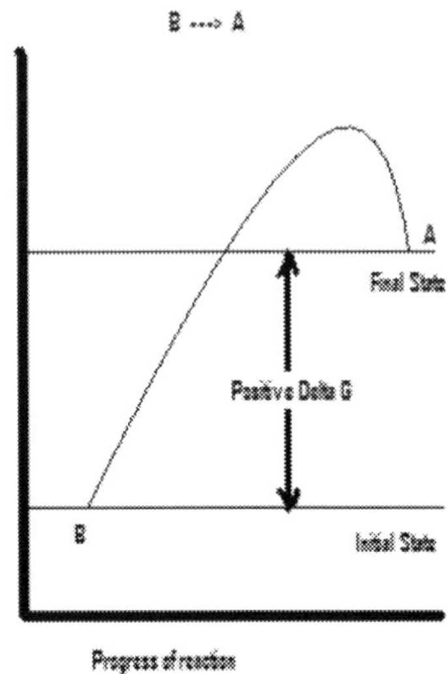
B ---> A
A
Final State
Positive Delta G
B
Initial State
Progress of reaction

Fig 11

ATP↓

Glucose →hexokinase Glucose 6 Phosphate

↕ ***phosphoglucose isomerase***

Fructose 6 Phosphate

ATP→↕ ***phosphofructokinase***

Fructose 1,6 bisphosphate

↓ aldolase

Glyceraldehyde 3 Phosphate

triose phosphate isomerase

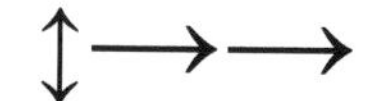

Dihydroxyacetone phosphate

Glyceraldehyde 3 phosphate dehydrogenase

1,3 bisphosphoglycerate

ATP←↕ ***phosphoglycerate kinase***

3 phosphoglycerate

↕ ***phosphoglycerate mutase***

2 phosphoglycerate

↕ ***Enolase***

Phosphoenolpyruvate

ATP←↕ ***pyruvate kinase***

Pyruvate ↔ Lactate

Fig 12

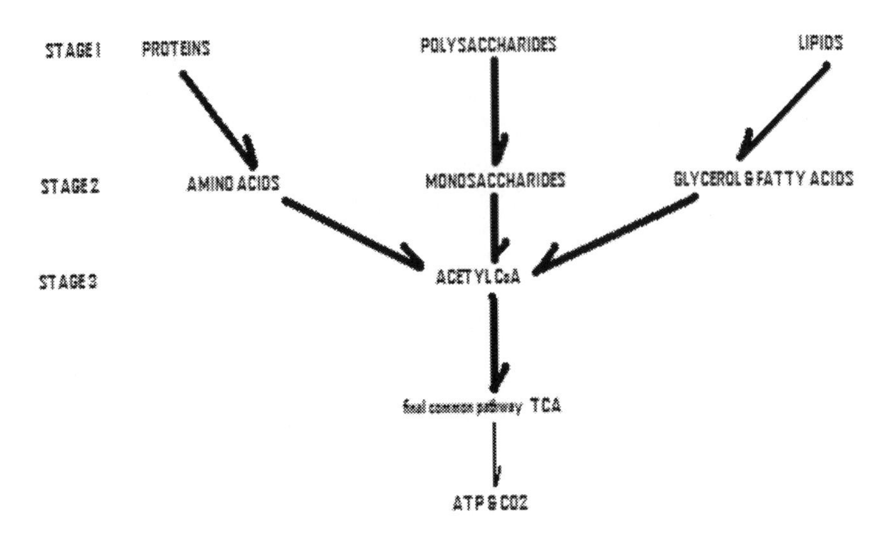

Stages of Catabolism

Fig 13

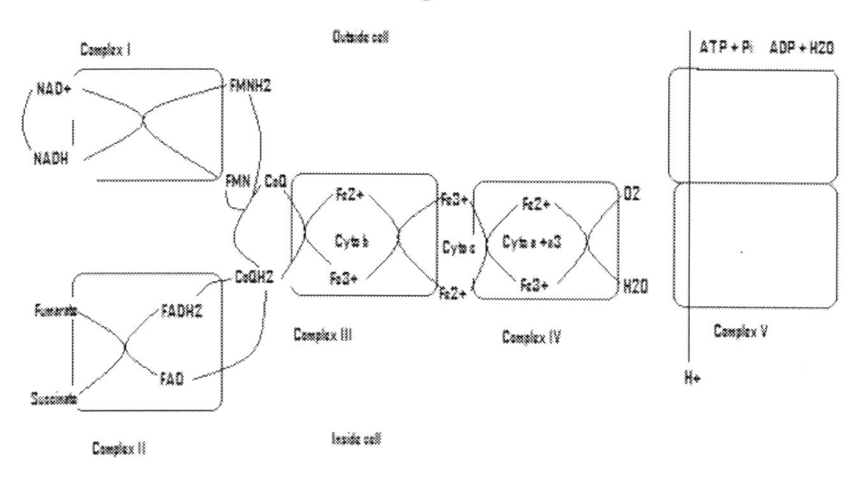

Electron Transport Chain Inner mitochondrial membrane

Fig 14

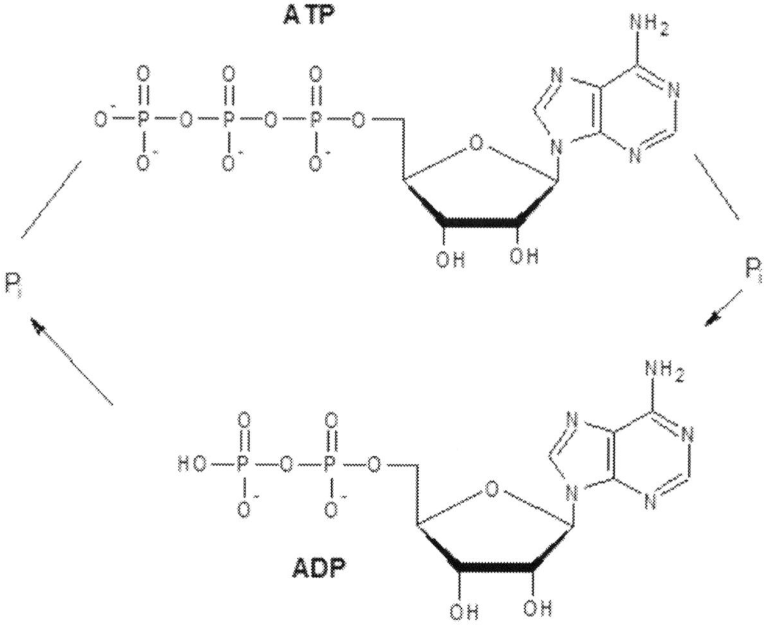

ATP

P_i

P_i

ADP

Fig 15

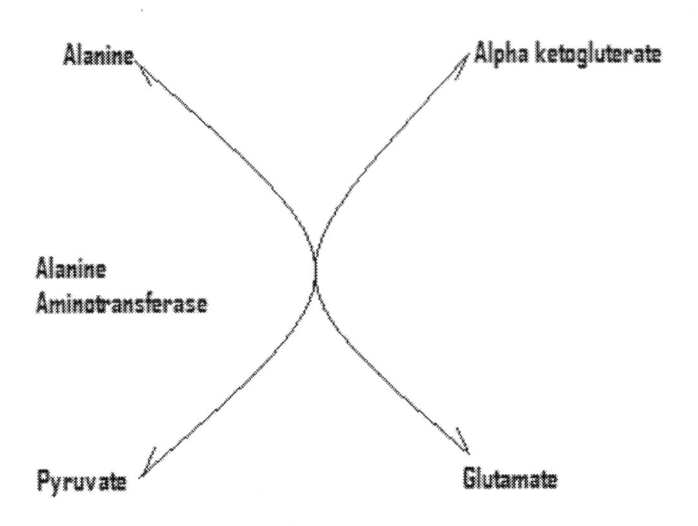

Fig 16

Glutamate

alpha ketogluterate

Pyridoxal
Phosphate

Pyridoxamine
Phosphate

Aspartate
Aminotransferase

Aspartate

Oxaloacetate

Fig 17

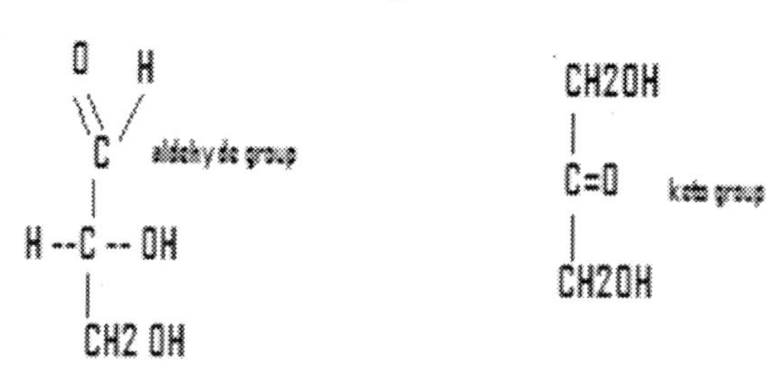

Fig 18

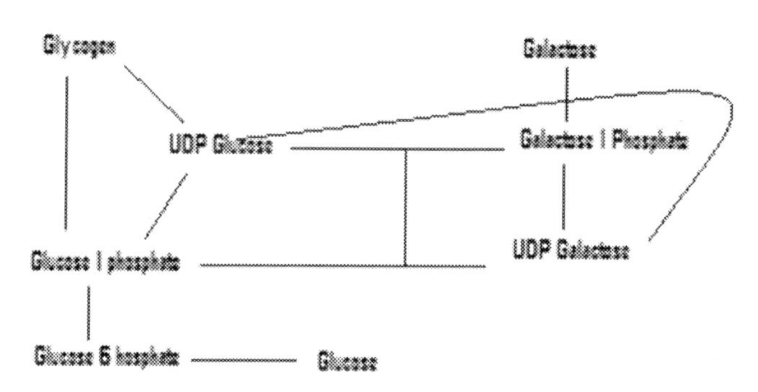

Fig 19

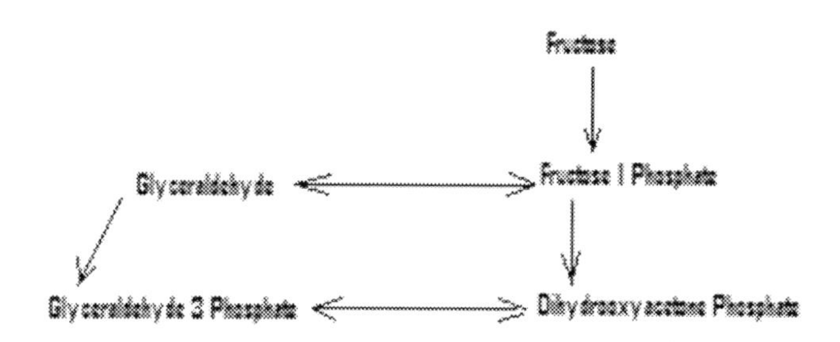

Fig 20

$$\text{Pyruvate} + \text{CoA} + \text{NAD}^{+} \rightarrow \text{acetyl CoA} + \text{CO}_2 + \text{NADH}$$

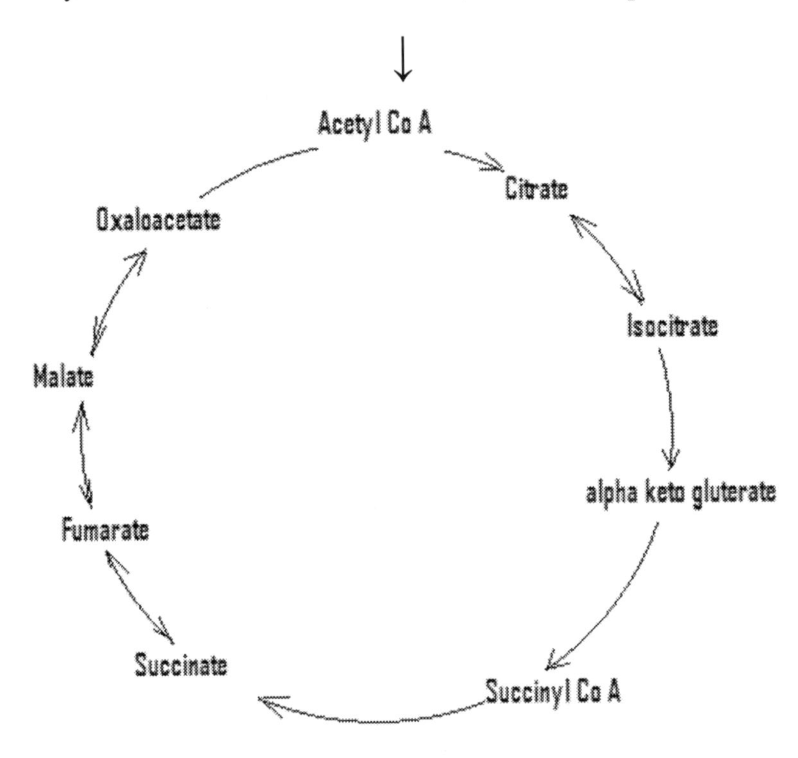

Fig 21

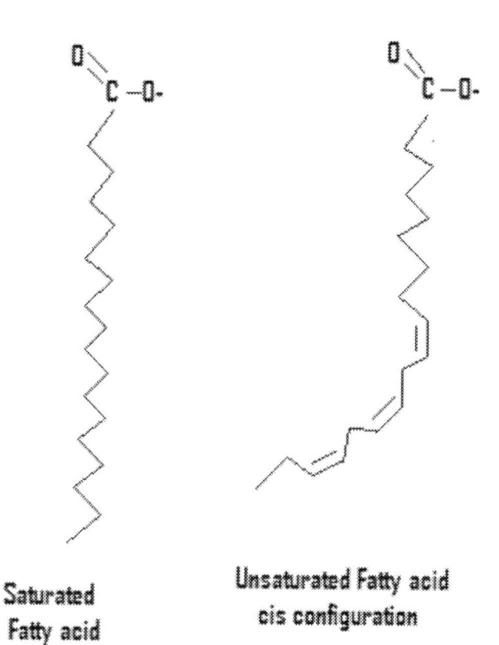

Fig 22

```
                          O
                          ||
        O      CH2 — O — C — R1
        ||      |
R2 — C — O — CH          O
                |          ||
               CH2 — O — C — R3
```

Triacylglycerol

Fig 23

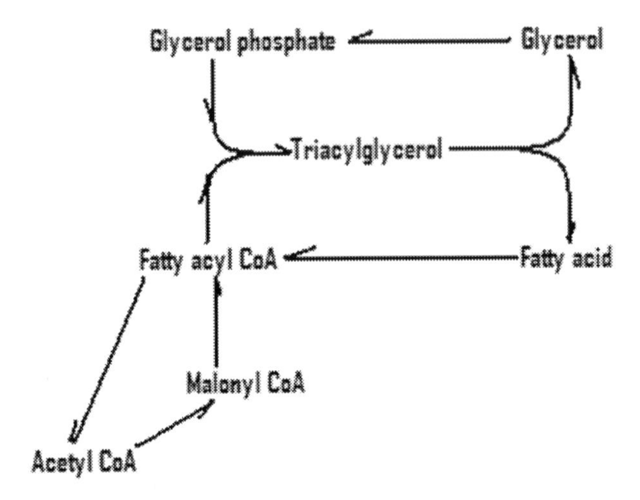

Triacylglycerol synthesis and degradation

Fig 24

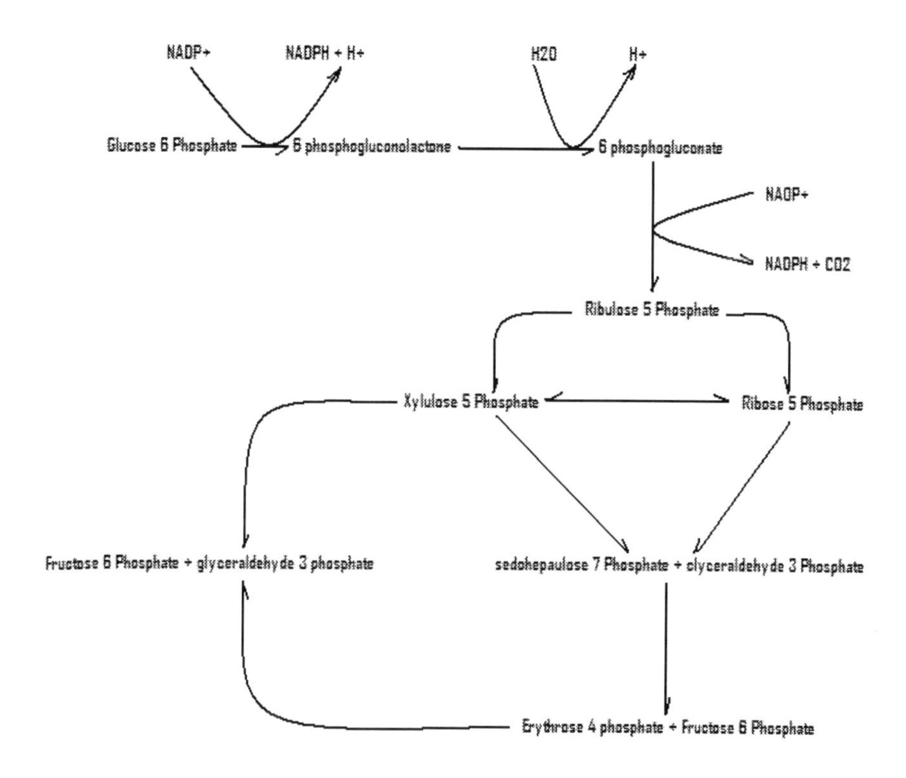

Fig 25

```
                               O
                               ||
        O          CH2— O — C — R1
        ||          |
R2 — C — O ——— C - H       O
                    |           ||
                   CH2 — O — P — O-
                                |
                               O-
```

Phosphatidic Acid

Fig 26

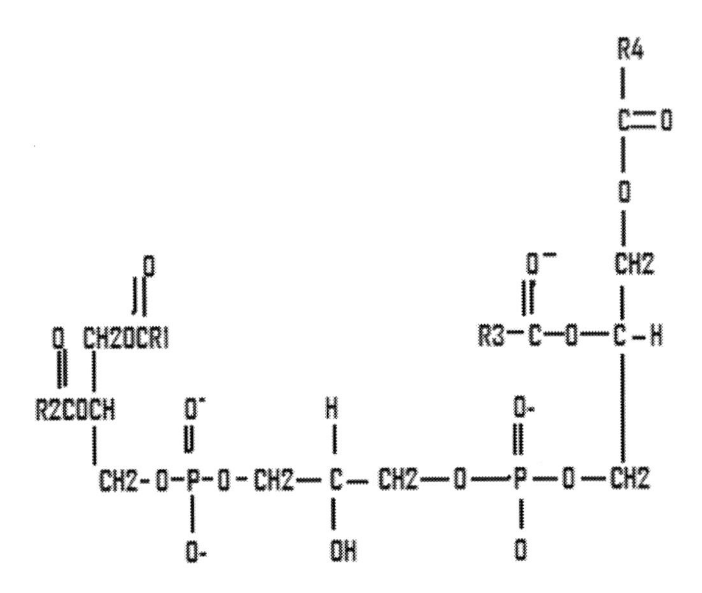

Cardiolipin

Fig 27

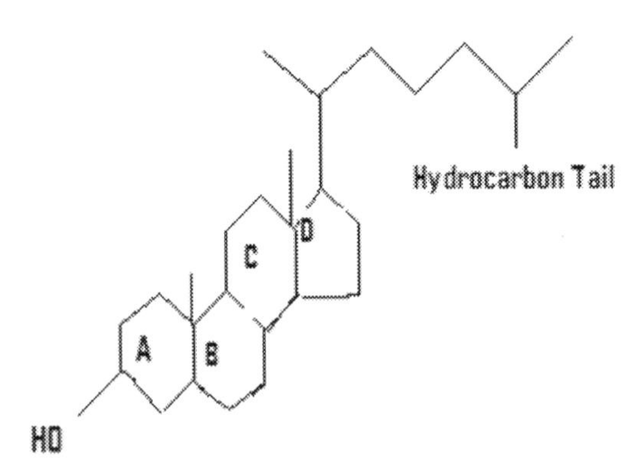

Fig 28

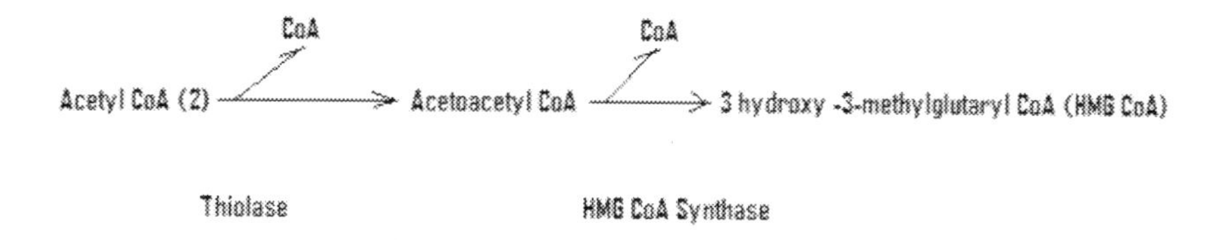

Fig 29

Percent composition	Very Low Density Lipoprotein	Low Density Lipoprotein	High Density Lipoprotein	Chylomicron
Triacylglycerol	60	8	5	90
Protein	5	20	40	2
Phospholipid	15	22	30	3
Cholesterol & Cholesteryl Esters	20	50	25	5

Fig 30

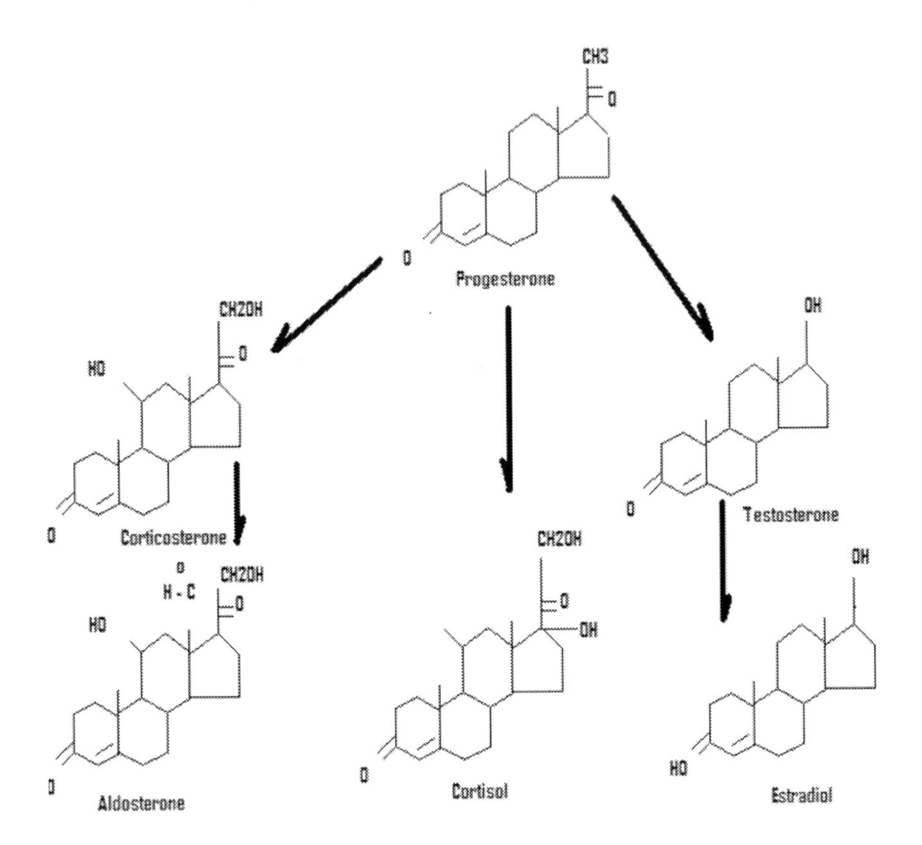
CH3
O
O
Progesterone
CH2OH
O
HO
O
Corticosterone
OH
O
Testosterone
CH2OH
O
OH
O
Cortisol
O
H - C
CH2OH
O
HO
O
Aldosterone
OH
HO
Estradiol

Fig 31

AMINO ACIDS	**GLUCOGENIC AMINO ACIDS**	**KETOGENIC AMINO ACIDS**	**BOTH KETOGENIC & GLUCOGENIC AMINO ACIDS**
NONESSENTIAL AMINO ACIDS	ALANINE ASPARGININE ASPARTATE CYSTEINE GLUTAMATE GLUTAMINE GLYCINE PROLINE SERINE		TYROSINE
ESSENTIAL AMINO ACIDS	ARGININE HISTIDINE METHIONINE THREONINE VALINE	LEUCINE LYSINE	ISOLEUCINE PHENYLALANINE TRYPTOPHAN

Fig 32

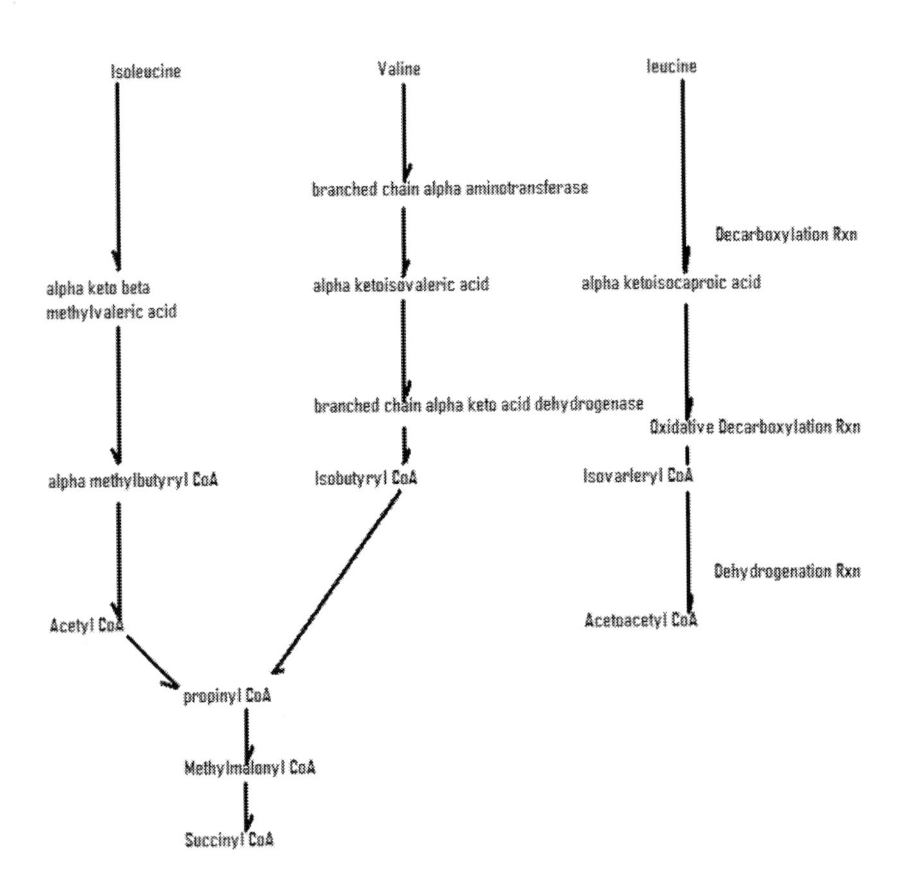
Isoleucine
Valine
leucine
branched chain alpha aminotransferase
Decarboxylation Rxn
alpha keto beta
methylvaleric acid
alpha ketoisovaleric acid
alpha ketoisocaproic acid
branched chain alpha keto acid dehydrogenase
Oxidative Decarboxylation Rxn
alpha methylbutyryl CoA
Isobutyryl CoA
Isovarleryl CoA
Dehydrogenation Rxn
Acetyl CoA
Acetoacetyl CoA
propinyl CoA
Methylmalonyl CoA
Succinyl CoA